United States
Environmental Protection
Agency

EPA 600/R-13/212 | November 2013 | www.epa.gov/ord

I0488631

Simulation Program i-SVOC
User's Guide

SCIENCE

Office of Research and Development

EPA/600/R-13/212

November 2013

Simulation Program i-SVOC

User's Guide

Software Version: 1.0

by

Zhishi Guo
Air Pollution Prevention and Control Division
National Risk Management Research Laboratory
U.S. EPA Office of Research and Development
Research Triangle Park, NC 27711, USA

National Risk Management Research Laboratory
Office of Research and Development
U.S. Environmental Protection Agency
Cincinnati, Ohio 45268

Disclaimer

The computer software described in this document was developed by the U.S. EPA for its own use and for specific applications. The Agency makes no warranties, either expressed or implied, regarding this computer software package, its merchantability, or its fitness for any particular purpose, and accepts no responsibility for its use. Mention of trade names and commercial products does not constitute endorsement or recommendation for use. The views expressed in this document are those of the author and do not necessarily represent the views or policies of the Agency.

Abstract

This document is the User's Guide for computer program i-SVOC, which estimates the emissions, transport, and sorption of semivolatile organic compounds (SVOCs) in the indoor environment as functions of time when a series of initial conditions is given. This program implements a framework for dynamic modeling of indoor SVOCs developed by the author, and covers six types of indoor compartments: air (gas phase), air (particle phase), sources, sinks (i.e., sorption by interior surfaces), contaminant barriers, and settled dust. Potential applications of this program include: (1) use as a stand-alone simulation program to obtain information that the current equilibrium models cannot provide, including evaluation of the effectiveness of certain pollution mitigation methods such as variable ventilation rates, source removal, and source encapsulation; (2) reducing the uncertainties in the existing multimedia models; and (3) use as a front-end component for stochastic exposure models to provide information about the SVOC distribution in indoor media in the absence of experimental data. This program is intended for advanced users, who are involved in and familiar with indoor environmental quality (IEQ) modeling or indoor exposure assessment. Because dynamic modeling of SVOCs in indoor media is a relatively new research field, a number of issues need to be resolved in future research. For example, efforts should be made to reduce the uncertainties in parameter estimation. There is a need to develop a data and knowledge base for key parameters for modeling indoor SVOCs, including, but not limited to, solid-air partition coefficient, solid-phase diffusion coefficient, and gas-phase mass transfer coefficient.

Acknowledgements

The author thanks the following testers for their contributions to the development of the simulation program and this User's Guide. Their feedback has enabled the author to locate and correct several errors in the code, remove a glitch in the setup program, and improve the usefulness and readability of this document.

External Testers:

Charles Weschler, Robert Wood Johnson Medical School and Rutgers University, USA
Doyun Won, National Research Council Canada
B. Beverly Guo and J. Jensen Zhang, Syracuse University, USA
Ying Xu, University of Texas at Austin, USA
John Little and Yaoxing Wu, Virginia Polytechnic Institute and State University, USA
Yinping Zhang, Tsinghua University, China
Xinke Wang, Xi'an Jiaotong University, China
Heidi Hubbard, ICF International, USA

EPA Internal Testers:

Kent Thomas, National Exposure Research Laboratory
Xiaoyu Liu, National Risk Management Research Laboratory
Christina Cinalli and Charles Bevington, Office of Pollution Prevention and Toxics
Laureen Burton, Office of Radiation and Indoor Air
Emmet Keveney, Region 2

Table of Contents

List of Tables

List of Figures

x

1. Introduction

1.1   What is i-SVOC?

I-SVOC is a Microsoft Windows-based computer program for dynamic modeling of the emissions, transport, and sorption of semivolatile organic compounds (SVOCs) in the indoor environment. This program implements a modeling framework recently developed by this author (Guo, 2013, 2014).

This program covers six types of indoor media (i.e., compartments): air (gas phase), air (particle phase), sources, sinks (i.e., sorption by interior surfaces), contaminant barriers, and settled dust. The key input parameters of the program include the solid-air partition coefficients, solid-phase diffusion coefficients, and gas-phase mass transfer coefficients.

Program i-SVOC is more of a *model shell* than a *model*. It can be regarded as a modeling platform on which the users can build and experiment with their own models. This program:

- Covers the functions of most commonly used models for indoor SVOC.
- Ensures that the calculated results are consistent with the existing mass transfer and empirical models.
- Provides the user with more flexibility than the existing models.
- Frees the user from numerical computations.

1.2   Main features

Over the past two decades, more than 20 mass transfer models and nearly 10 multimedia models have been developed for indoor VOCs and SVOCs (Liu, et al., 2013; Guo, 2013; and references herein). While the mass transfer models have contributed significantly to better understanding of the behavior of diffusional sources, sinks, and barriers in buildings, their applications in the real world have been somewhat limited because of two major issues: model incompatibility and computational complexity (Guo, 2013). As a result, it is difficult to use those models to simulate the cases where multiple sources and sinks are present. Particulate matter plays an important role in human exposure to SVOCs. Most existing multimedia models assume that there is an instantaneous equilibrium between the gas-phase and particle-phase SVOCs. It has been demonstrated that such assumption tends to overestimate the particle-phase SVOC concentrations in room air when the particle-air partition coefficient is large (Guo, 2014).

1

Program i-SVOC resolves the above-mentioned problems by adopting a new modeling framework, as described in Section 6. The key features of this program include:

- It is the first general-purpose tool for dynamic modeling of indoor SVOCs.
- It allows multiple diffusional sources and sinks to be present in a simulation.
- It includes components for dynamic modeling of SVOC interactions with suspended and settled particles, which can be size-segregated.
- It frees the user from tedious numerical computations and allows them to focus on developing models and conducting simulations.
- There are multiple options for the simulation output, including SVOC concentrations in indoor media and mass fluxes.

Details about the specifications of this program are given in Tables 11 and 12 in Chapter 5.

1.3 Potential applications

Program i-SVOC may be useful in the following areas:

- It can be used as a stand-alone simulation program to obtain information that the current equilibrium models cannot provide. For example, i-SVOC can evaluate the effectiveness of certain pollution remediation methods such as variable ventilation rates, source removal, and source encapsulation.
- It can help reduce the uncertainties in the existing multimedia models. For instance, for the SVOCs with large solid-air partition coefficients, the instantaneous equilibrium assumption tends to overestimate the particle-phase SVOC concentrations in room air. Program i-SVOC can provide an estimate of the degree of sorption saturation (DSS), which can be used as an adjusting factor.
- It can be used as a front-end component for stochastic exposure models, such as the EPA's Stochastic Human Exposure and Dose Simulation (SHEDS) model (http://www.epa.gov/heasd/research/sheds.html). Program i-SVOC can provide estimates of the SVOC distributions in indoor media in the absence of experimental measurements.

1.4 Intended users

Program i-SVOC is mainly for advanced users who are involved in and familiar with indoor environmental quality (IEQ) modeling or indoor exposure assessment.

Many resources are available for those who would like to learn more about IEQ modeling. For example, a general introduction of IEQ modeling is given by Sparks (2001); commonly used parameter estimation methods for IEQ modeling are reviewed by Guo (2002); methods for modeling indoor particles are described by Nazaroff (2004).

2

1.5   Limitations

Dynamic modeling of SVOC distributions in buildings is a relatively new research field. There are many unresolved issues. Potential users should be aware of the limitations of the current program.

- The uncertainties in the simulation results are directly linked to the uncertainties in the model parameters such as the partition and diffusion coefficients. There have been many discussions in the literature on how to estimate the partition coefficients. On the other hand, data and models for the solid-phase diffusion coefficient remain scarce. While there are well-established methods for estimating the gas-phase mass transfer coefficient for flat surfaces, there is great uncertainty in this parameter associated with airborne and settled particles. Thus, reducing the uncertainties in parameter estimation is a key for future research.
- The current version of program i-SVOC is for a single air zone and a single SVOC. The program does not support chemical reactions.
- Program i-SVOC ignores the interactions between the settled dust and the surface which the dust particles are in contact with.
- Direct contact of settled dust with a source is an important mass transfer mechanism by which SVOCs migrate from the substrate into the dust. The current version of i-SVOC cannot simulate this mass transfer process. A simple method for estimating the upper bound of SVOC migration is given in Appendix A.
- Many indoor pesticides are SVOCs. This program does not contain any mass transfer models for liquid applications.
- Program i-SVOC does not perform stochastic simulations.
- Program i-SVOC is mainly a research tool. It is not intended for regulatory purposes.

1.6   Hints on parameter estimation

1.6.1 Parameter estimation is the user's responsibility

As a general-purpose simulation tool for indoor SVOCs, program i-SVOC and this User's Guide do not provide any default values for the conditions that the users want to simulate, or recommend any specific methods for parameter estimation. It is the user's responsibility to select proper parameter values. Among the many parameters used in i-SVOC, solid-air partition coefficient, solid-phase diffusion coefficient, and gas-phase mass transfer coefficient are critical. There is a vast pool of literature that discusses the methods for estimating these parameters. To further improve our ability to model indoor SVOCs, it is a logical next step to compile and evaluate the existing parameter estimation methods and, if necessary, develop new methods. Very brief discussions on estimating the above-mentioned parameters are provided below.

1.6.2 Solid-air partition coefficient

Many methods are available for estimating the solid-air partition coefficient for interior surface materials and particles. It can be estimated from either the vapor pressure or octanol-air partition coefficient of the chemical. It appears that more researchers are in favor of using the latter (for example, Finizio et al., 1997; Weschler & Nazaroff, 2010). Although experimentally determined partition coefficients are scarce, some data are available for certain SVOCs (For example, Xu & Little, 2006; Guo et al., 2012).

1.6.3 Solid-phase diffusion coefficient

Solid-phase diffusion coefficient is difficult to predict because it depends on not only the properties of the chemical but also those of the substrate. This parameter is commonly related to the molecular weight of the chemical (Schwope, et al., 1990 and the references herein). Correlations for the chemicals within the same class are well established (Guo, 2002 and references herein). Limited experimental data are available for this parameter (For example, Schwope et al., 1990; Xu & Little, 2006; Guo et al, 2012).

1.6.4 Gas-phase mass transfer coefficient

Several methods are available for estimating the gas-phase mass transfer coefficient for materials with flat surfaces (e.g., building materials and furniture). Computer program PARAMS (Guo, 2005) implements three methods, among which the method based on Sherwood number is most commonly used. The information needed to compute this parameter includes the formula of the chemical, air velocity (or speed), temperature in the room, and surface area.

Methods for estimating the gas-phase mass transfer coefficient for airborne particles are also available (For example, Li & Davis, 1996). However, there is great uncertainty in this parameter. The cited values for indoor particles differ by several orders of magnitude. Furthermore, little information is available about the range of gas-phase mass transfer coefficient for settled dust. More research is needed to reduce the uncertainty associated with this parameter.

2. Software Installation and Technical Support

2.1   Hardware and software requirements

Program i-SVOC runs on personal computers with Microsoft Windows as the operating system. The program has been tested for Windows XP, 7, and 8. A minimum of 5 MB free disk space is required. The screen resolution should be at least 1024 by 768 pixels. Internet connection is required only for downloading the installation package from the EPA website.

2.2   Installing program i-SVOC

If your computer is connected to a local network, it is likely that you need the "Administrative Privileges" to install this program.

After the setup program is downloaded, use the File Explorer to locate the file. Double click the file name and then follow instructions. The default target folder for installation is "C:\Program Files\EPA_ISVOC\" for Windows XP and "C:\Program Files (x86)\EPA_ISVOC\" for Windows 7 and 8. During the installation, the setup program will create an icon on your desktop screen. For Windows 8, a "tile" for the application will also be created on the start screen. Click the icon or tile to start the simulation program. You can also start the program from the Programs menu.

During the installation, the User's Guide will be copied to the target folder. It can be accessed from the Programs menu. There is also an icon on your desktop screen.

The setup program has been scanned by anti-virus software Symantech Endpoint Protection (Symantech, Mountain View, CA) and is free of known viruses.

2.3   Uninstalling program i-SVOC

For Windows XP or 7, go to Windows Control Panel, select <Add or remove programs>, select i-SVOC from the program list, and then click <Change/Remove>.

For Windows 8, right-click the <Desktop> tile from the start screen and then click <All apps>, which is located at the bottom-right corner of the screen. This will bring up a list of apps. Click <Control Panel> and then click <Programs/Uninstall a program>. In the next window, select

<Programs and Features>, <Uninstall a program>. Select program i-SVOC from the list and then click <Uninstall/Change>.

2.4 Contact information

If you have any questions or would like to report bugs and errors, please contact the developer:

Zhishi Guo
US EPA Office of Research and Development
National Risk Management Research Laboratory
Air Pollution Prevention and Control Division
Indoor Environment Management Branch
Mail Code E305-03
Research Triangle Park, NC 27711, USA
Telephone: +919-541-0185
Email: guo.zhishi@epa.gov

2.5 Technical support

Limited technical support is provided by the developer. Communications by electronic mails are highly preferred. If your problem is associated with a particular model, please attach the model file to your message.

3. Getting Started

3.1   User interface

3.1.1 General appearance

This simulation program features a multi-page design. As shown in Figure 1, there is a menu bar on top. Nine "speed buttons" below the menu bar provide quick access to frequently used menu items. Table 1 lists the names of the speed buttons and their corresponding menu items. If you move the cursor over a speed button, a hint will be displayed momentarily.

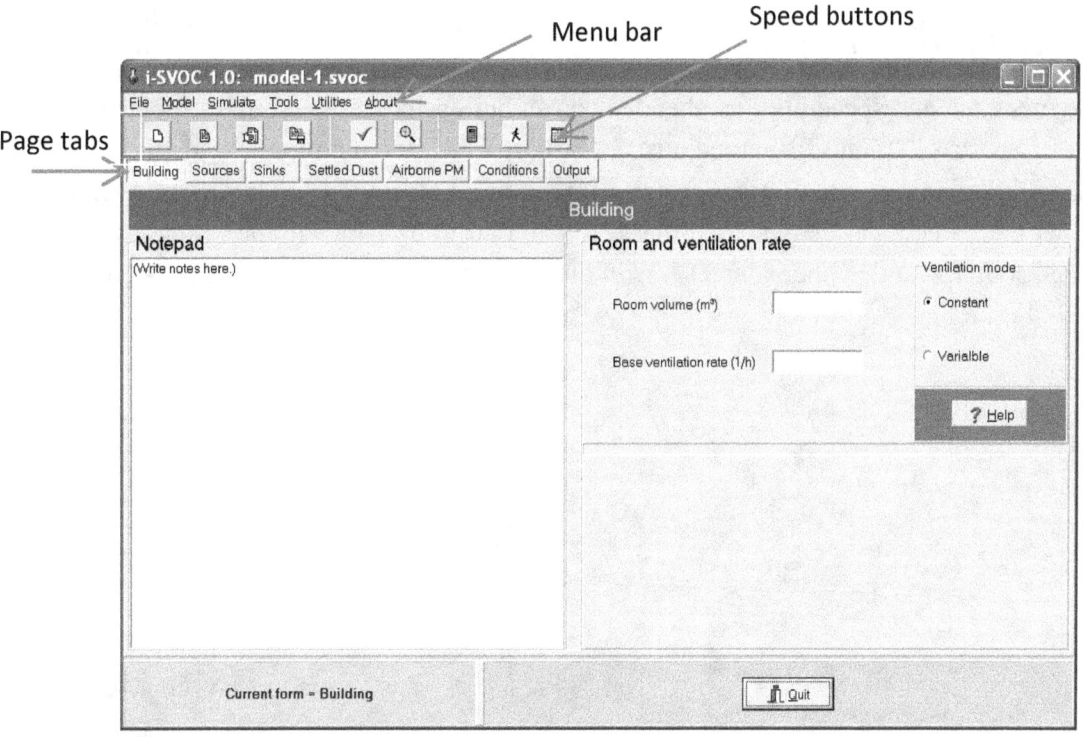

Figure 1. User interface of program i-SVOC: (1) parameter entry page for building properties.

Table 1. Speed buttons and their corresponding menu items.

Button Group	Position (from left)	Button Name	Corresponding Menu Item	Hint Text
File manager	1	New	File/New	Create a new model
	2	Open	File/Open	Open an existing model
	3	Open recent	File/Open recent file	Open a recent model
	4	Save	File/Save	Save current model
Error checking	5	Compile	Model/Compile	Compile current model
	6	Inspect	Model/Inspect	Inspect current model
Simulation mode	7	Run	Simulate/Run	Run model at normal speed
	8	Run slow	Simulate/Run slow	Run model at reduced speed
	9	Run & save	Simulate/Run & save	Run model & put back output [a]

[a] Described in Section 4.7.

3.1.2 Pages and forms for user input

Below the menu bar and speed buttons, there are seven button-like tabs, and they are labeled "Building," "Sources," "Sinks," "Settled dust," "Airborne PM," "Conditions," and "Output" (Figure 1). All the input parameters are grouped into the first six pages. For example, to enter a source, click the "Sources" tab. Note that there are two data entry forms in the <Sources> page: "Diffusional sources" and "Other sources," with the form tabs located near the lower-left corner, as shown in Figure 2.

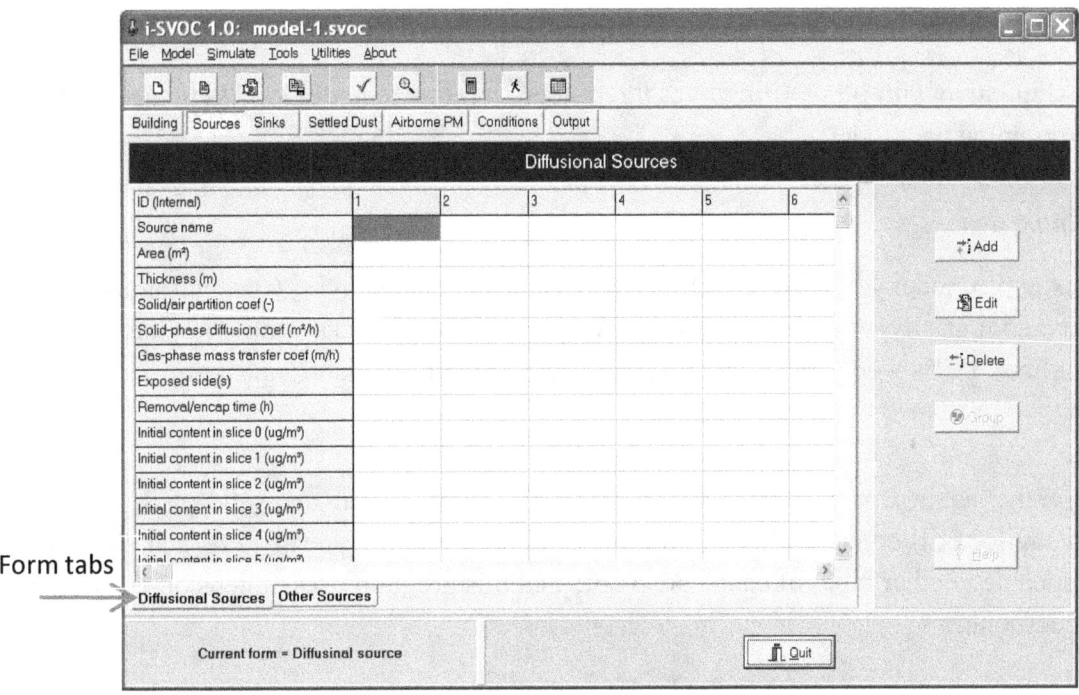

Figure 2. User interface of program i-SVOC: (2) parameter entry form for diffusional sources. Note the two form tabs near the lower-left corner.

3.1.3 Output tables

The simulation results are posted to the <Output> page, in which there are two data tables, one for SVOC concentrations and the other for SVOC fluxes. Click the <Output> page tab to access those tables.

Program i-SVOC does not create graphics. You can use either the <Copy> or <Copy All> button to transfer the results to a spreadsheet and then work from there. The difference between the two commands is as follows: the <Copy> command copies only the area that you highlighted while the <Copy all> command copies everything in the table without highlighting.

3.1.4 Display adjustment

Depending on the version of the operating system and the language packs installed, some computers cannot display superscript Arabic numbers correctly. For example, the unit for diffusion coefficient, (m^2/h), may appear as (m?h) on screen. To resolve this problem, click <Utilities> / <Change display> from the main menu, and then click <OK>. After this change, the unit will be displayed in regular font, as (m2/h). You can switch the display mode any time.

3.2 Creating a model for a diffusional source

Making a simulation with i-SVOC involves three steps: creating a model, checking the model for errors, and running the model.

3.2.1 Creating a model

The first example model you will create is for polychlorinated biphenyl (PCB) emissions from a caulking material in a hypothetical room with a volume of 100 m^3 and a ventilation rate of 1 air change per hour. PCBs were once used as a plasticizer for caulk, sealants, and paint. These applications were banned in the U.S. in 1978.

Table 2 lists the required parameters. The solid-air partition coefficient and solid-phase diffusion coefficient are for PCB congener 52. These values are for a specific type of caulk and, thus, may not be applicable to other types of caulk. Also note that this program uses dimensionless solid-air partition coefficient.

Table 2. Example parameters for modeling PCB-52 emissions from a caulking material.

Page/Form	Parameter Name	Value	Notes
Building	Room volume (m^3)	100	
	Ventilation rate (h^{-1})	1	
Sources/ Diffusional sources	Source name	Caulk	
	Exposed area (m^2)	0.2	
	Thickness of source material (m)	0.01	
	Solid-air partition coefficient (dimensionless)	6.54×10^7	[a]
	Solid-phase diffusion coefficient (m^2/h)	2.25×10^{-11}	[a]
	Gas-phase mass transfer coefficient (m/h)	2.2	[b]
	Initial concentration in the source (μg/m^3)	8.07×10^9	[a][c][d]
	Exposed side(s)	top	
Simulation conditions	Initial concentration in room air (μg/m^3)	0	
	Simulation duration (h)	10000	
	Output data points	200	
	Output options	--	[e]

[a] Data from Guo et al. (2011), page 104. [b] Estimated by computer program PARAMS (Guo, 2005).
[c] Assume the concentration in the source is uniform initially. [d] Pay attention to the units; use equation 41 in Section 6.7.1 to convert (μg/g) to (μg/m^3); there is a unit converter under the <Tools> menu.
[e] Explained in the text below Figure 4.

To get started, run i-SVOC. After the title window appears, click <Go>. Now your screen should look like Figure 1 in Section 3.1.1. There is a notepad on the left side for making notes. It is always a good idea to describe your model in detail for future reference. Try to type a few words there, such as "This is my first model…" Next, enter the room volume and ventilation rate. Now you are done with the first data entry form.

To enter the parameters for the diffusional source, click the <Sources> page tab. If the current form is not for diffusional sources, click the <Diffusional sources> tab at the lower-left corner. Now your screen should look like Figure 2 in Section 3.1.2.

To add a source, click the <Add> button. A data entry form will be displayed (Figure 3). This program requires that each source be given a name, any name except a blank. This requirement may seem superfluous. However, its importance will become apparent as the source name is the only information that can help you identify the data columns in the output table.

Now enter the required parameters on the left side of the form. Program i-SVOC accepts numerical values in either general or exponential formats:

CORRECT:	3.15	1256	1.23E8	1.23e8	3.45E-12
INCORRECT:	1,256	1.23*10^8			

There is a check box near the bottom for "Source removal." Ignore it at this moment. It will be explained later in Sections 3.4.2 and 4.2.

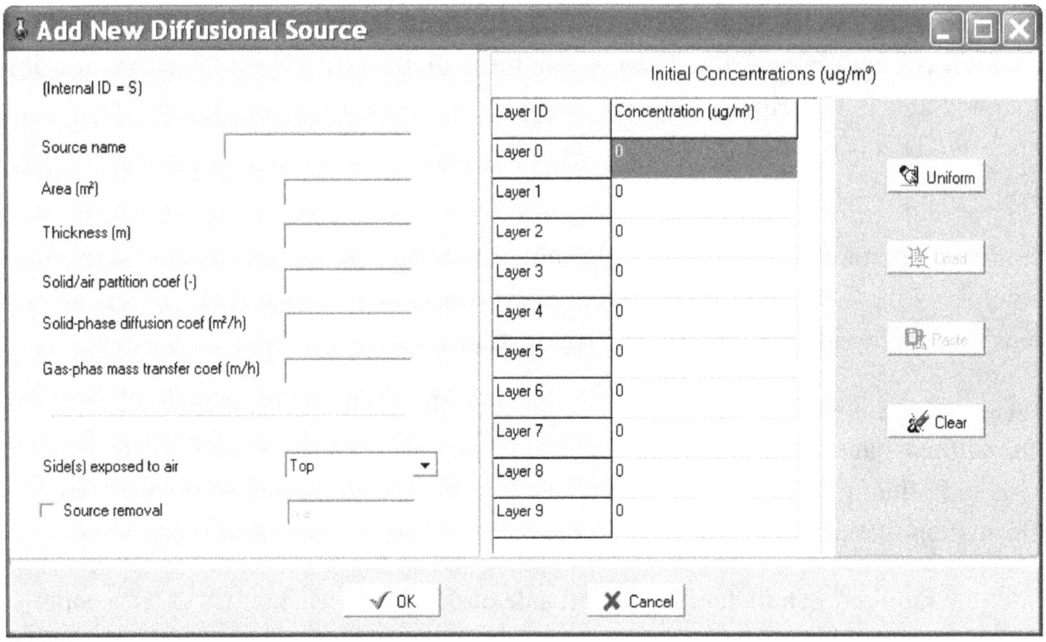

Figure 3. Data entry form for diffusional sources.

Next, move to the right side of the form to enter the initial concentrations in the source. Remember, i-SVOC is based on the modified state-space (MSS) method, which divides the source into a finite number of slices (Guo, 2013). In this program, a source made of the same material (i.e., a single layer source) is divided into ten slices. Program i-SVOC does not require that the initial concentrations in the source be uniform. Because we have assumed that the concentration is uniform initially (see Table 2), all the layers will have the same concentration value (i.e., 8.07×10^9 $\mu g/m^3$). If you do not want to type in 10 identical numbers, there is a shortcut: click on the first cell, type in "8.07E9," and then click the <Uniform> button.

After all is done, click the <OK> button. If an error message pops up, fix the problem and then click <OK>. This program allows the user to enter multiple sources. To add another source, click the <Add> button again.

You can make changes to the parameters that you have already entered in the form. To do so, click the source name first, and then click the <Edit> button. To delete a source, click the souece name first, and then click <Delete>.

To enter the parameters for simulation conditions, click the <Conditions> page tab. As shown in Figure 4, the user is asked to enter three parameters on the left side of the form: initial concentration in air, simulation duration, and the number of data points to be provided in the output table. Enter the values according to Table 2.

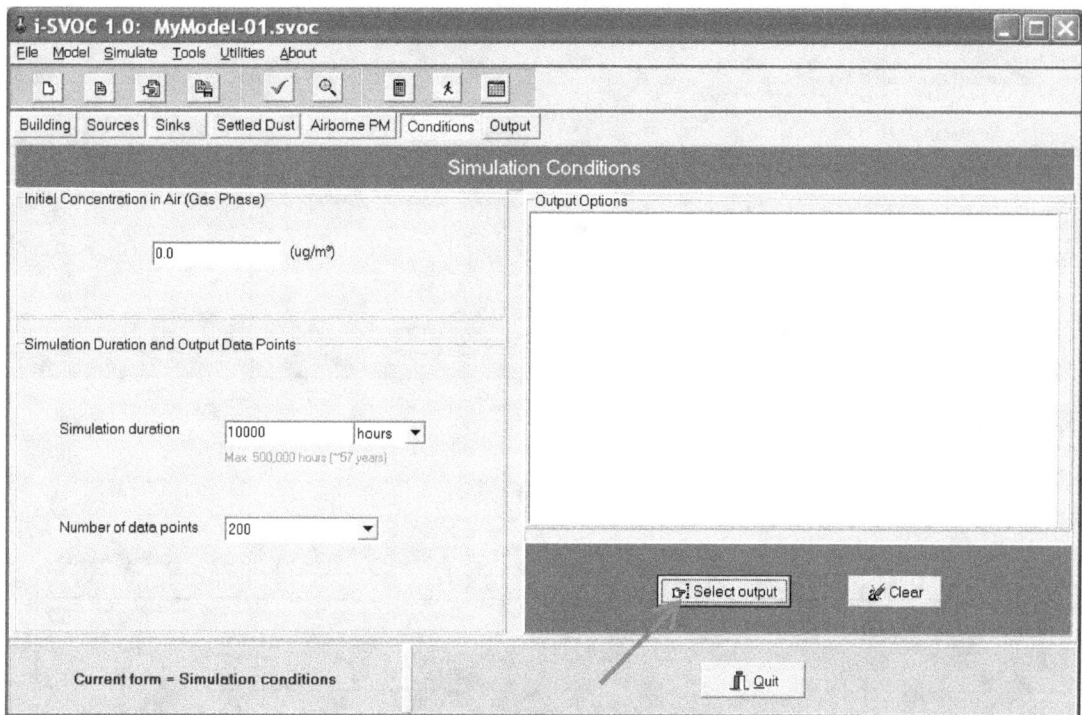

Figure 4. The <Conditions> page. Click the <Select output> button to choose output options.

As the last step, you need to select the output data types. As you will see in Section 5, program i-SVOC can generate different types of output data and, consequently, the data table can be very large. Here you are given the opportunity to determine the types of output data you need. To select output options, click the <Select output> button. As shown in Figure 5, the options are listed on the left pane. Select the items you need and then click the <Select> button. For your first model, select the following:

    00) Air: gas-phase only ($\mu g/m^3$)

Click the <OK> button after you are done. The <Conditions> form should look like Figure 6. Now you have completed your first model with i-SVOC.

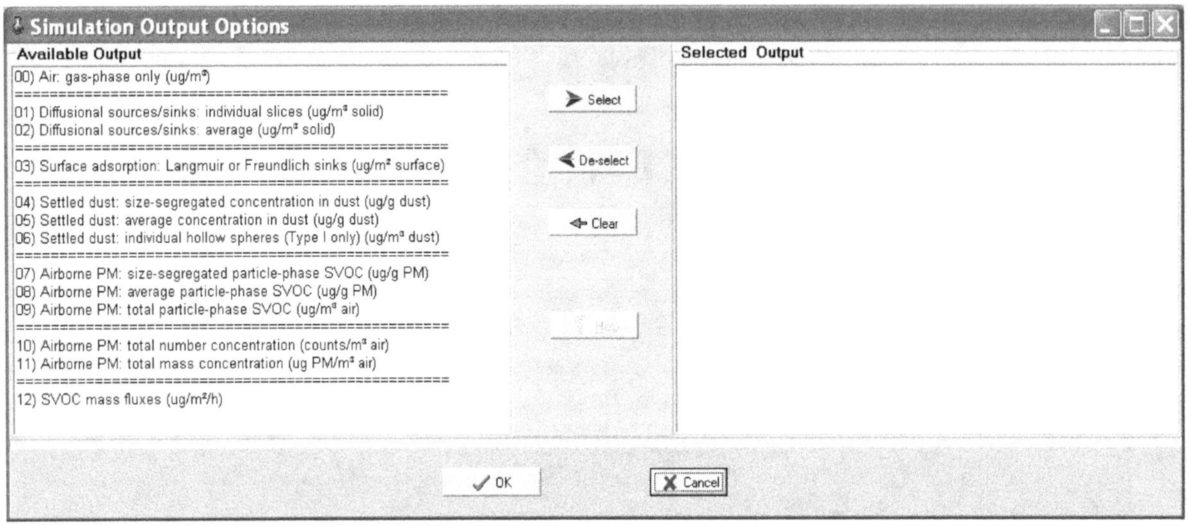

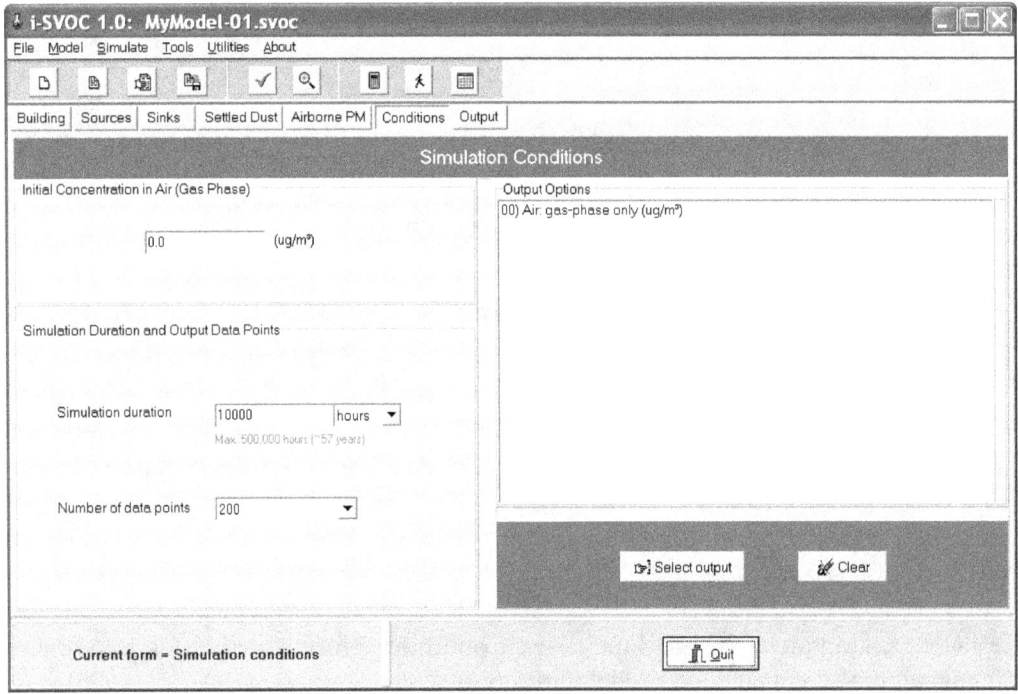

Figure 5. Form for selecting the output data types.

Figure 6. Completed form for simulation conditions.

Before moving any further, let's save the model for future use. To save the current model, click the <Save> speed button (the fourth from left) or select <File> / <Save> from the main menu. Name the file "MyModel-01" and save it to a folder. Note that i-SVOC uses file extension ".svoc" for model files.

3.2.2 Error checking

The next step is to check the input parameters for potential errors. Program i-SVOC provides two ways for error checking: <Compile> and <Inspect>. The former allows i-SVOC to read your model and check for potential errors; the latter allows the user to check for errors.

To compile the current model, click the <Compile> speed button (the one with a red check sign) or select <Model> / <Compile> from the main menu. Try it now. If you see an error message, fix the problem and try again.

To check for errors by yourself, click the <Inspect> speed button (the one with a magnifying glass) or select <Model> / <Inspect> from the main menu. Program i-SVOC will generate a laundry list for the parameters in the current model. It tells you how the program interprets your model. It is highly recommended that you go over the list to confirm the correctness of the parameters. The parameter list can be printed.

3.2.3 Running the model

To start the simulation, click the <Run> speed button (the one with a calculator glyph) or select <Simulate> / <Run> from the main menu. If you entered the parameters correctly, there should be no problem during the simulation.

The simulation results are posted to the first data table (the one with the "Concentrations" tab) located in the <Output> page. As mentioned in Section 3.1.2, i-SVOC does not create graphics. You can use either the <Copy> or <Copy All> button to transfer the results to a spreadsheet and then work from there. Use the <copy> command to copy only the area that you highlighted, or use the <Copy all> command to copy everything in the table without highlighting. Figure 7 shows the air concentration profile created with a spreadsheet program.

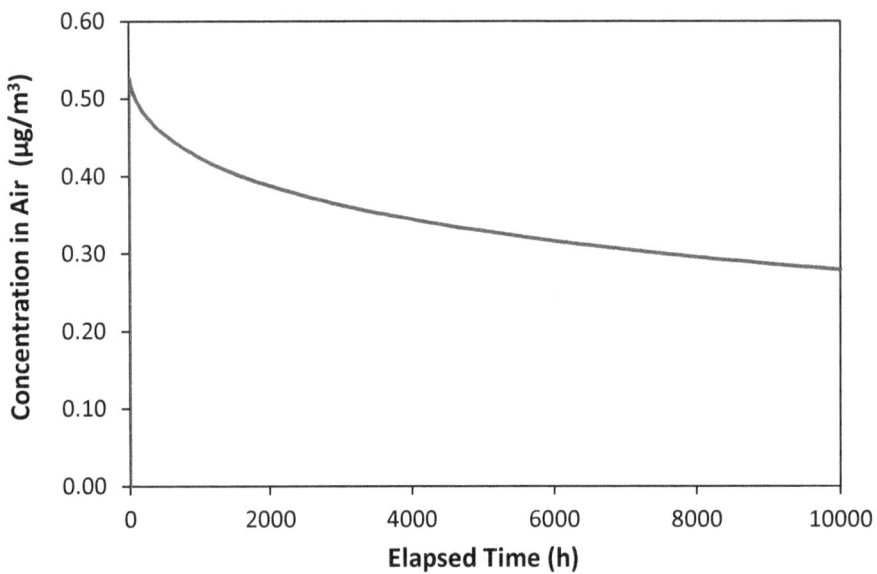

Figure 7. Simulated PCB-52 concentrations in room air (model MyModel-01.svoc).

3.2.4 More output options

This program can do more than just calculating the air concentrations. Go to the <Conditions> page, click the <Select output> button, and then add the following output option:

01) Diffusive sources/sinks: individual slices ($\mu g/m^3$ solid)

Re-run the model and you will see ten more data columns in the output table. They are for SVOC concentrations in individual slices in the source. Scroll all the way down, you will find the name of the source (or sink), slice ID, and thickness and depth for each slice. With this information, you can use a spreadsheet program to create a plot similar to Figure 8. Note that the thicknesses of the slices are not equal. They are determined by the modified state-space method. See Section 6.3.1 below and Guo (2013) for more details.

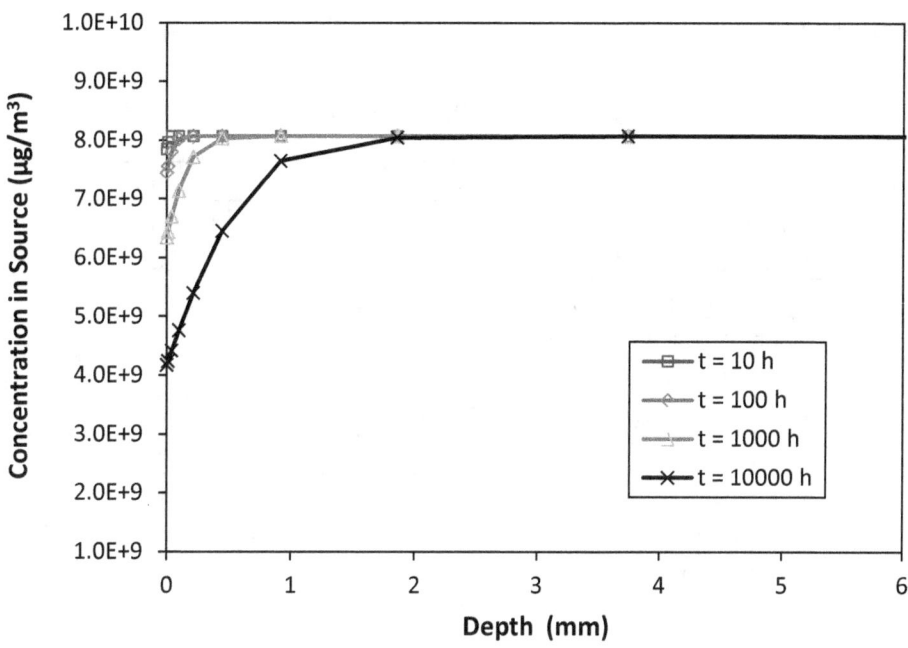

Figure 8. Concentration profiles for PCB-52 in the source as functions of depth and time. The exposed surface is at depth = 0.

3.2.5 Where are the temperature and air velocity in this model?

Temperature and air velocity (or air speed) are both important parameters for mass transfer. Why aren't they listed in Table 2? This program does not require these parameters explicitly because they are imbedded in other key parameters. For example, room temperature is reflected in the partition, diffusion, and mass transfer coefficients. Similarly, air velocity is imbedded in the gas-phase mass transfer coefficient.

3.3   Adding a diffusional sink

3.3.1 Example sink parameters

Sorption of SVOCs by interior surfaces is an important issue in both exposure assessment and contamination remediation. Program i-SVOC allows simulations with multiple sinks. Let's add a diffusional sink to the model that you have just created. The sink is the concrete walls with a total area of 80 m^2. The new parameters are given in Table 3.

Table 3. Example parameters for concrete walls as a diffusional sink for PCB-52.

Page/Form	Parameter Name	Value	Notes
Sinks/ Diffusional sinks	Sink name	Concrete	
	Exposed area (m^2)	80	
	Thickness of sink material (m)	0.025	
	Solid-air partition coefficient (dimensionless)	2.11×10^7	[a]
	Solid-phase diffusion coefficient (m^2/h)	2.98×10^{-11}	[a]
	Gas-phase mass transfer coefficient (m/h)	0.68	[b]
	Initial concentration in the sink (μg/m^3)	0	
	Exposed side(s)	top	

[a] From Guo et al. (2012), page 73. [b] Estimated by computer program PARAMS (Guo, 2005) by assuming the average area of each wall is 20 m^2.

3.3.2 Open an existing model file

In this practice, we will use file MyModel-01.svoc that you created earlier. There are two methods to open an existing model file: (1) Click the <Open> speed button (the second from left) or by selecting <File> / <Open> from the main menu. (2) If the file was saved recently, click the <Open recent> speed button (i.e., the third from the left), or select <File> / <Open recent file> from the main menu.

3.3.3 Create and run the new model

Click the <Sinks> page tab. This page contains two forms: <Diffusional sinks> and <Langmuir and Freundlich sinks>. Go to the former. The data entry form for diffusional sinks is almost identical to the one for diffusional sources (i.e., Figure 3). Enter the parameter values provided in Table 3. Then, click the <OK> button.

Now compile and then inspect your model just as you did for your first model. Next, save the new model to file "MyModel-02" for future use by clicking <File>, <Save as> from the main menu. Note that, if you click <File>, <Save> or the <Save> speed button, the new model will be saved to "MyModel-01". Now, run the model and you will get the results shown in Figures 9 and 10. Note that the travel distance for SVOCs in a sink is rather short, and that the average concentration in the sink depends on the thickness selected. To obtain the concentration profiles like Figure 10, a proper thickness should be selected by trial and error.

18

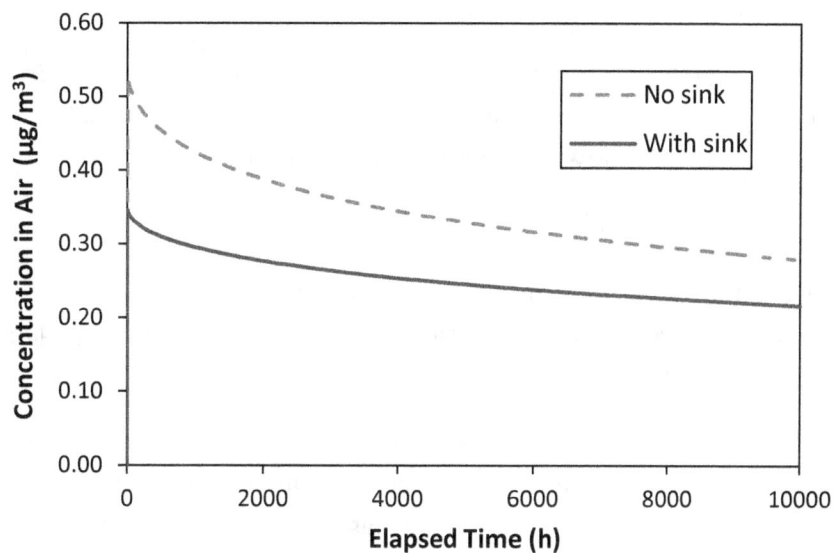

Figure 9. Simulated PCB-52 concentration in room air by using models MyModel-01.svoc (no sink) and MyModel-02.svoc (with sink).

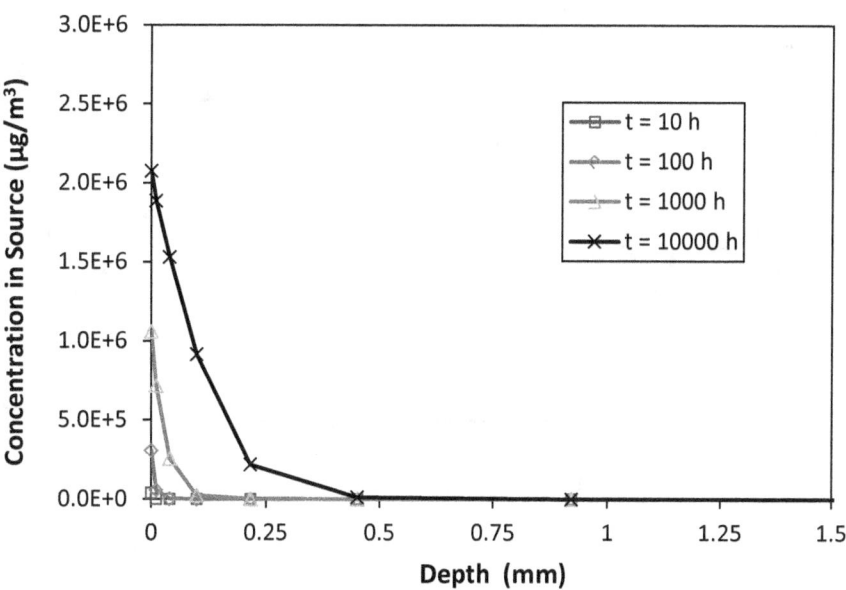

Figure 10. Concentration profiles in the sink substrate as functions of depth and time. The exposed surface is at depth = 0.

3.3.4 Include mass fluxes in the output options

In addition to calculating the SVOC concentrations in indoor media, i-SVOC can also calculate SVOC fluxes (i.e., emission or sorption rates) between indoor media. To include SVOC fluxes in the simulation output, add the following item to the output options:

12) SVOC mass fluxes ($\mu g/m^2/h$)

Re-run the model and the mass flux data will be posted to the second data table. Note that mass fluxes can be either positive or negative. *A positive value means emission and a negative value sorption* (Figures 11 and 12).

Note that the distinction between a diffusional source and a diffusional sink is surperficial. A source can become a sink and vice versa. It all depends of the fugacity difference between the source (or, sink) and room air. In other words, the mass transfer models treat the diffusional sources and sinks the same (Kumar and Little, 2003a).

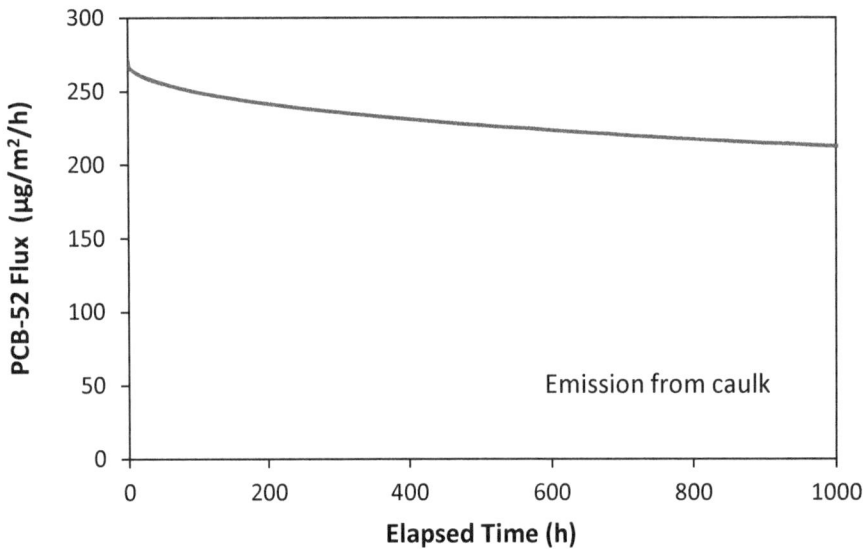

Figure 11. PCB-52 flux between the source (caulk) and room air; positive values mean emissions from source.

20

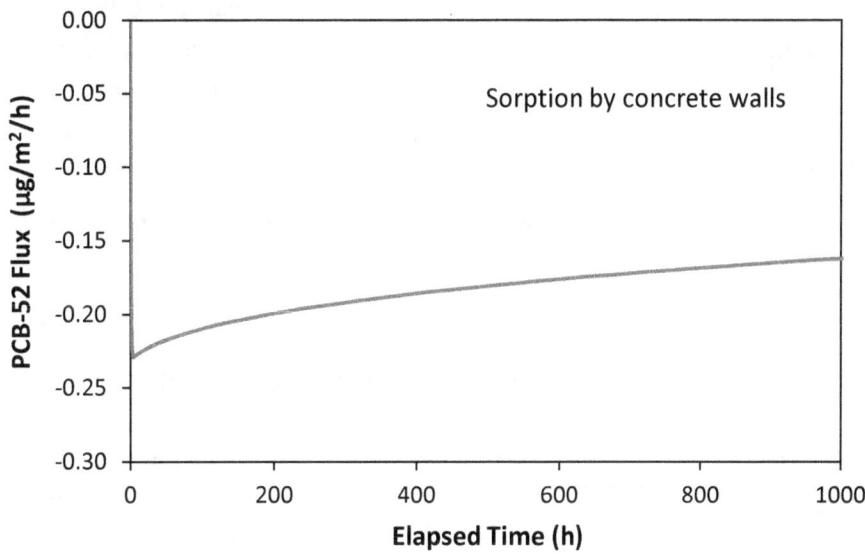

Figure 12. PCB-52 flux between for the sink (concrete walls) and room air; negative values mean sorption from air.

3.4   Creating a model for settled dust

Program i-SVOC can simulate both permeable and impermeable settled dust (Guo, 2014). For each category, multiple size bins and dust types are allowed, as shown in the example below. The dust models are fully compatible with the source and sink models. Table 4 lists the example parameters for sorption of di-ethyl-phthalate (DEP) by settled dust.

3.4.1 Creating a constant emission source

Program i-SVOC includes several non-diffusional source models (i.e., empirical source models) and they serve for two purposes:

- They are useful for modeling the data from test chambers; and
- They can be used to simplify the model, especially in the cases where airborne or settled particles are involved.

In this practice, we will use an empirical emission model to keep the gas-phase DEP concentration at 1 $\mu g/m^3$ in the absence of any sinks. To start from scratch, click the <New> speed button (i.e., the first from left) or select <File>, <New> from the main menu. Complete the <Building> page by entering 30 m^3 for the room volume and 1 h^{-1} for the base ventilation rate.

Then, go to the <Sources> page. Click the <Other sources> tab near the lower left corner. Click <Add> to bring up the data entry form (Figure 13). Click on the dropdown list at the top-left corner and then select "Constant source (rate)". The form will display the model description and the required parameters (Figure 14). Give the source a name and then enter 30 μg/h for the emission rate. In the absence of any sinks, this value gives a steady-state DEP concentration of 1 μg/m^3 for a 30-m^3 room with a ventilation rate of one air change per hour. Ignore the "Shut-off" time at this moment. Click <OK>.

Table 4. Example parameters for simulating DEP sorption by settled dust with diameters of 2.5, 10, and 50 μm.

Page/Form	Parameter Name	Value		
Building	Room volume (m^3)	30		
	Ventilation rate (h^{-1})	1		
Other sources	Source type	Constant source (rate)		
	Emission rate (μg/h)	30		
Dust Type I (Permeable)	Dust name	D2.5	D10	D50
	Diameter (μm)	2.5	10	50
	Particle-air partition coefficient (dimensionless) [a]	1.24×10^8	1.24×10^8	1.24×10^8
	Particle-phase diffusion coefficient (m^2/h) [b]	1×10^{-13}	1×10^{-13}	1×10^{-13}
	Gas-phase mass transfer coefficient (m/h) [c]	5	4.5	4
	Initial SVOC concentration in particles (μg/g)	0	0	0
	Dust particle density (g/cm^3)	1	1	1
	Number of dust particles [d]	2.44×10^{11}	3.82×10^9	3.06×10^7
Simulation conditions	Initial concentration in air (μg/m^3)	0		
	Simulation duration (h)	100		
	Number of output data points	100		
	Output options	[e]		

[a] Average value based on two estimation methods, from Weschler et al. (2008).
[b] Rough estimate based on several references.
[c] Best guesses.
[d] Assuming a dust loading of 0.2 g/m^2 for each size bin and a floor area of 10 m^2. A unit converter is available under the <Tools> menu.
[e] The output option is: 04) Settled dust: size-segregated concentration in dust (μg/g dust).

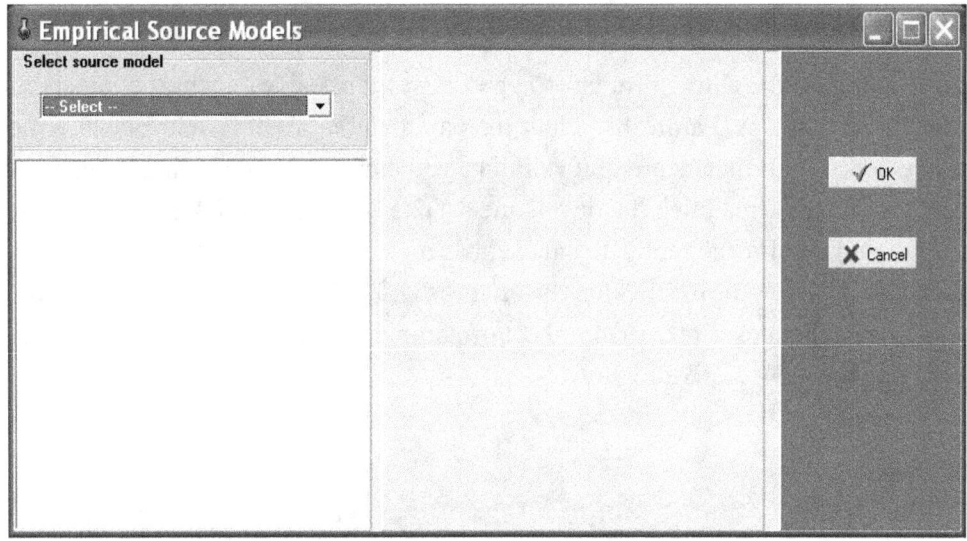

Figure 13. Parameter entry form for empirical source models before the model is selected.

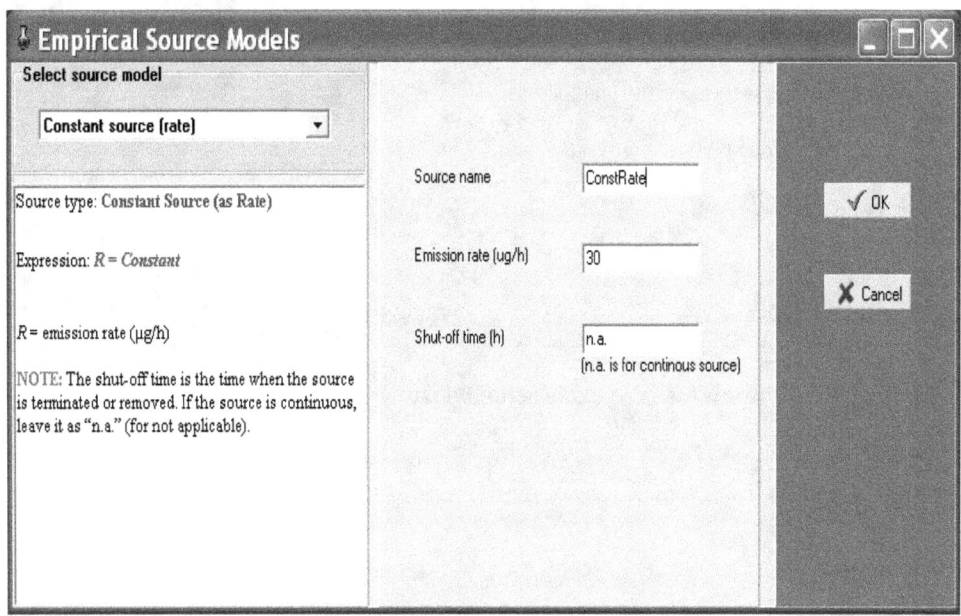

Figure 14. Parameter entry form for empirical source models after the model is selected and parameters are entered.

3.4.2 SVOC accumulation in settled dust

Go to page <Settled dust>. Click form tab <Type I dust (permeable)> near the lower-left corner. In the parameter entry form (Figure 15), enter the parameters for D2.5 first. Because the remaining two sets of parameters are quite similar, you can save some typing by using the <Replicate> command. First, click the dust name ("D2.5") and then click the <Replicate> button. Note that the dust name for the replicate data set has an extra "x" because i-SVOC does not allow replicate names in a form. Click on dust name "D.25x" and then click the <Edit> button. Now you can make changes. Enter the third set of parameters in a similar manner. After you finish, the screen should look like Figure 16.

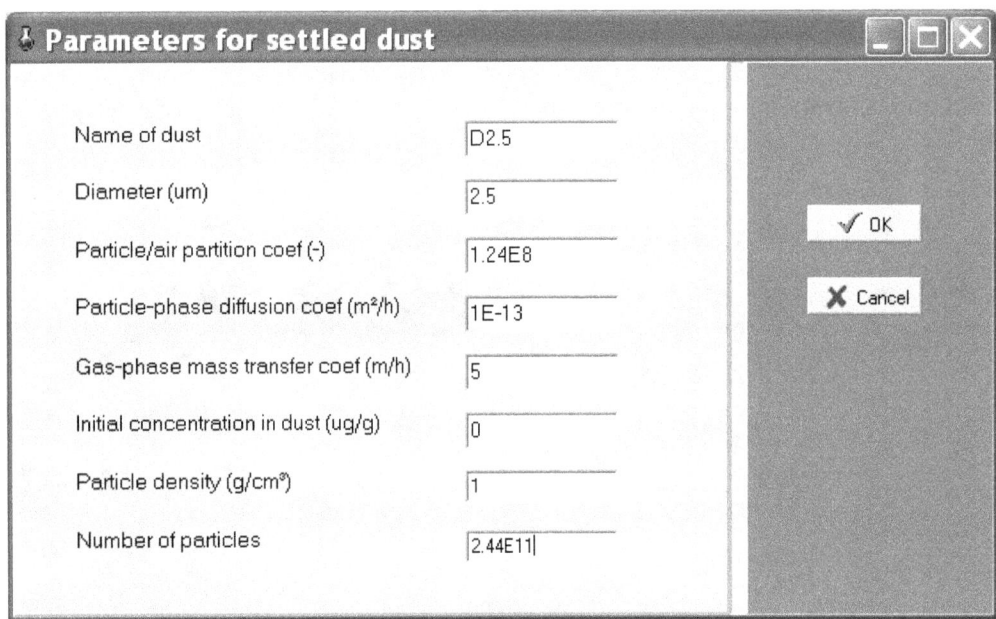

Figure 15. Parameter entry form for permeable settled dust.

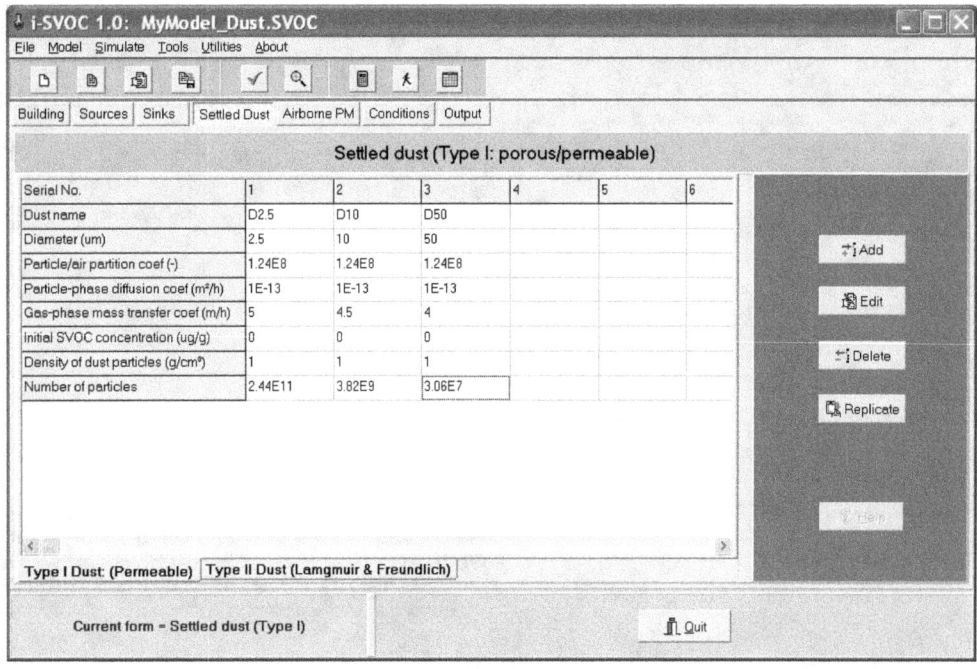

Figure 16. The form for type I dust with input parameters for three size bins.

In the <Conditions> page, enter the remaining parameters. The completed form should look like Figure 17.

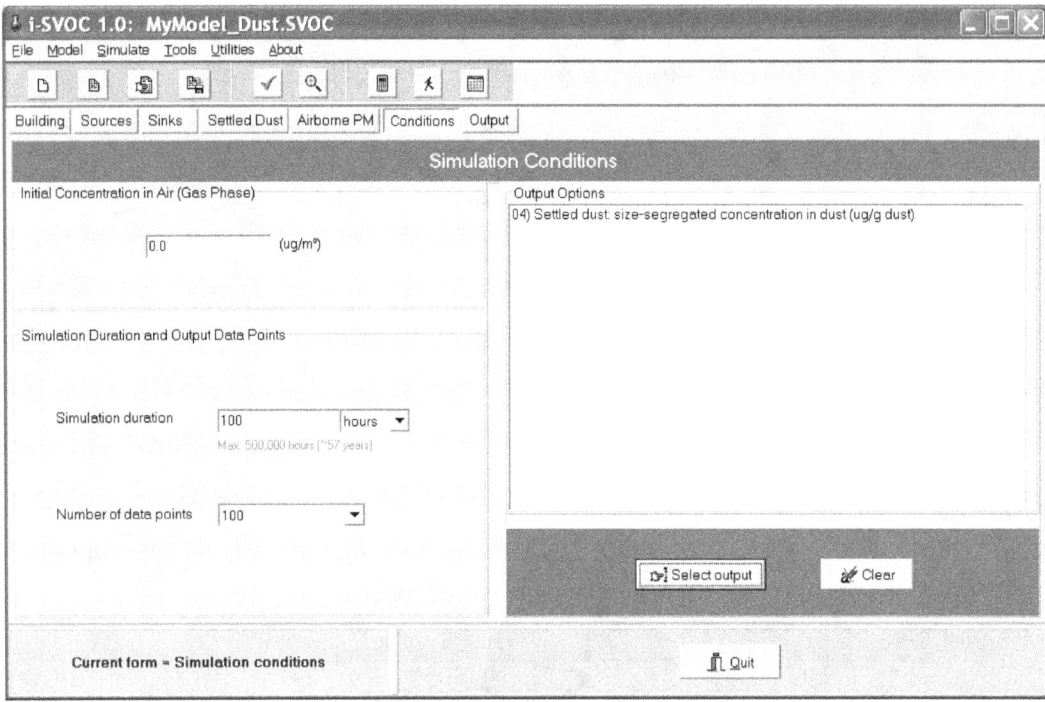

Figure 17. Completed form for simulation conditions.

Now save the model to external file "MyModel_dust.svoc". Next, <Compile> and then <Inspect> the model. When you are ready, click the <Run> speed button. In case the simulation fails, try the <Run slow> option (the second speed button from right). The simulation results are shown in Figure 18.

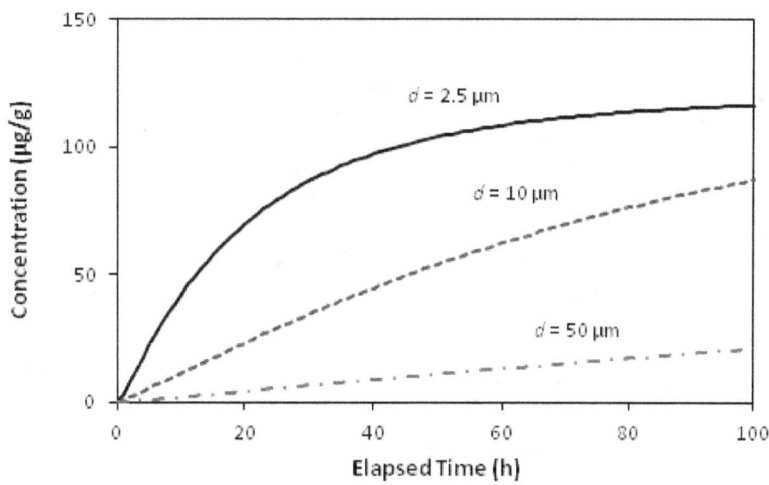

Figure 18. Simulation DEP concentrations in settled dust with three size bins.

3.4.3 SVOC re-emission from settled dust after the source is removed

Let's make a slight modification to the model you have just created. Go to the <Other sources> form, click on the source name, and then click <Edit>. In the data entry form, the shut-off time is "n.a." (for not applicable). Change it to 100. Click <OK>. Then go to the <Conditions> page. Change the simulation duration to 200 hours. Re-compile the model and click the <Run> speed button. The results are shown in Figure 19.

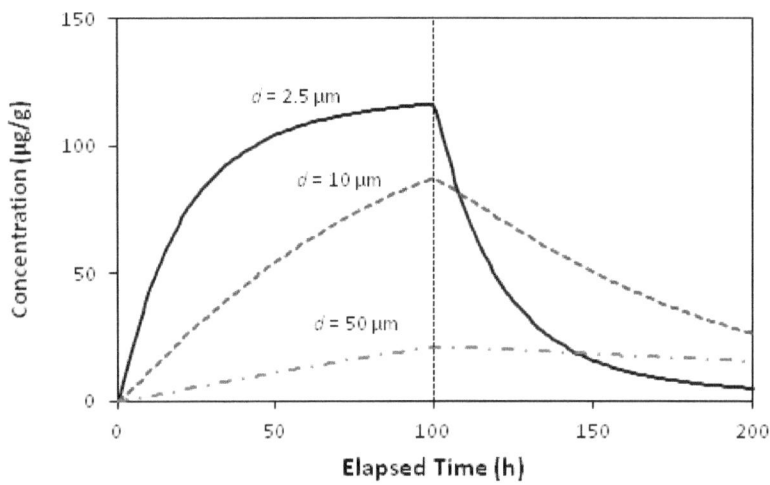

Figure 19. Simulated DEP losses from settled dust after the source is removed.

3.5   Creating a model for airborne particles

Under typical indoor environmental conditions, airborne particles can stay suspended for only a short period of time. The *residence time* is usually less than an hour (Qian et al., 2008; Guo, 2014). Typically, the simulation duration for airborne particles should be no longer than several hours. Simulation results for airborne particles without consideration of residence time can be misleading. In most cases, long-term simulations are useful only for studying the SVOC interactions with individual particles, not for the particle "population".

In this practice, we will use the parameters listed in Table 5 to investigate the DEP sorption by airborne particles in four size bins. Go to the <Building> to enter the room volume and ventilation rate. Next, go to the <Other sources> under the <Sources> page to define the source.

Table 5. Example parameters for simulating DEP sorption by airborne particles with diameters of 0.1, 1, 2.5, and 10 µm.

Page/Form	Parameter Name	Value			
Building	Room volume (m³)	30			
	Ventilation rate (h⁻¹)	1			
Sources/ Other sources	Source type	Constant source (rate)			
	Emission rate (µg/h)	30			
	Shut-off time (h)	(Leave it as is)			
Airborne PM/ Type I (permeable)	Name of airborne particles	APM0.1	APM1	APM2.5	APM10
	Particle release mode	Constant	Constant	Constant	Constant
	Diameter (µm)	0.1	1	2.5	10
	Number concentration (counts/m³) [a]	3.82×10^{10}	3.82×10^{7}	2.45×10^{6}	3.82×10^{4}
	Initial SVOC concentration in particles (µg/g)	0	0	0	0
	Particle-air partition coefficient (dimensionless) [b]	1.24×10^{8}	1.24×10^{8}	1.24×10^{8}	1.24×10^{8}
	Particle-phase diffusion coefficient (m²/h) [c]	1×10^{-13}	1×10^{-13}	1×10^{-13}	1×10^{-13}
	Particle density (g/cm³)	1	1	1	1
	Gas-phase mass transfer coefficient (m/h) [d]	2000	1000	500	300
	Deposition rate constant (h⁻¹) [e]	0.45	0.45	0.6	0.75
	Initial concentration in air (µg/m³)	1			
Simulation conditions	Simulation duration (h)	2			
	Number of output data points	100			
	Output options	See text below			

[a] Each of these values is equivalent to a mass concentration of 20 µg/m³. Thus, the total particle concentration in air is 60 µg/m³.
[b] Average value based on two estimation methods, from Weschler et al. (2008).
[c] Rough estimate based on several references.
[d] Best guesses based on Shi & Zhao (2012).
[e] Rough estimates based on Qian et al. (2008).

29

Airborne particles are defined in the <Airborne PM> page, which contains three forms for, respectively, permeable particles, impermeable particles (i.e., Langmuir and Freundlich adsorption), and solid particles covered with a liquid film. Go to the first form and then click the <Add> button. Enter the parameters in Table 5. The completed form should look like Figure 20.

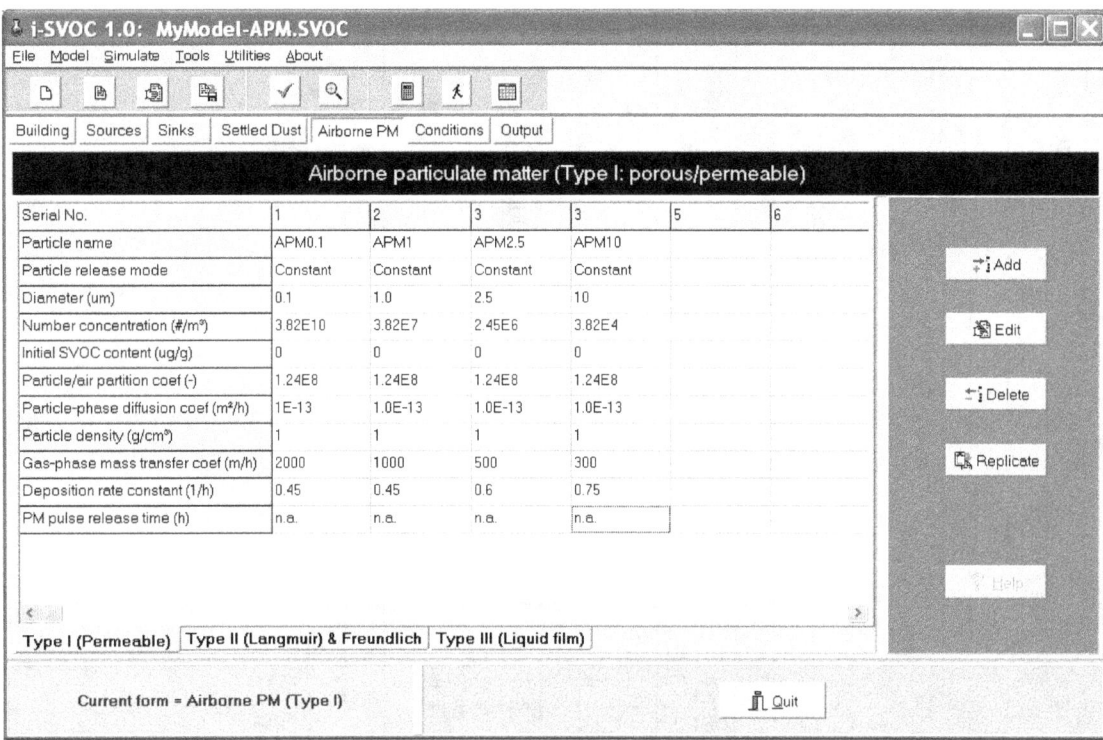

Figure 20. Completed parameter entry form for type I airborne particles.

In the <Conditions> page, enter 1 for the initial air concentration, 2 for simulation duration, and 100 for the number of output data points. For the output option, select

07) Airborne PM: size-segregated particle-phase SVOC (µg/g PM).

Save the model as "MyModel-APM.svoc." After error-checking, click the <Run slow> speed button (the second speed button from right). The difference between <Run> and <Run slow> is discussed in Section 3.7 below.

CAUTION: The simulation results for airborne particles should be interpreted with discretion. The results in Figure 20 are the time history for individual particles. *For a given particle size or type, the SVOC concentration at the residence time is representative of the particle "population"*

30

if the particle concentration remains constant during the simulation (Guo, 2014). Particles with different sizes may have different deposition rates and, thus, slightly different residence times. The value shown in Figure 21 is for 2.5 μm-diameter particles.

CAUTION: It is also recommended that, when simulating airborne particles with constant release rates, the gas-phase SVOC concentration remain constant.

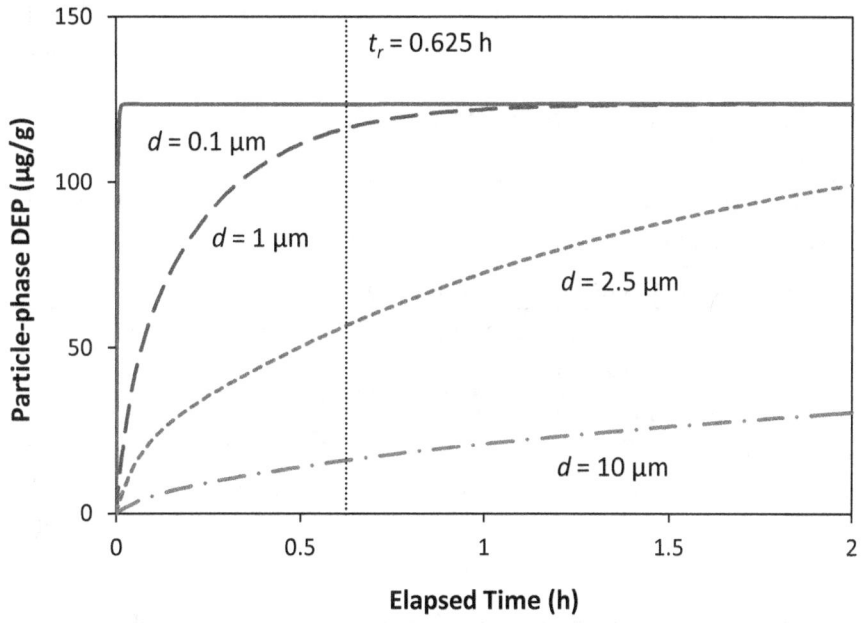

Figure 21. Simulated DEP concentrations in airborne particles as a function of time. The residence time (t_r) shown is for 2.5 μm-diameter particles.

3.6   Modeling phthalate emissions from vinyl flooring

Until now, we have dealt with hypothetical cases. In the next example, we will use the parameters in the literature to predict the concentrations of di-2-ethlhexyl phthalate (DEHP) in two types of test chambers, known as CLIMPAQ (Chamber for Laboratory Investigations of Materials, Pollution and Air Quality) and FLEC (Field and Laboratory Emission Cell). The data were generated by Clausen et al. (2004) and the model parameters by Xu and Little (2006). We will create a model to calculate the chamber concentrations with different loading factors.

3.6.1 Parameters

The parameters for the CLIMPAQ chamber data reported by Xu and Little (2006) are summarized in Table 6. Note that the unit for time has changed from second to hour.

Table 6. Parameters for DEHP emissions from vinyl flooring (from Xu et al., 2006).

Parameter Group	Parameter Name and Units	Symbol	Parameter Value	
			CLIMPAQ	**FLEC**
Chamber	Chamber volume (m^3)	V	0.051	3.5×10^{-5}
	Ventilation rate (h^{-1})	--	10	780
	Initial concentration in source ($\mu g/m^3$)	C_0	2.6×10^{11}	2.6×10^{11}
Source	Source area (m^2)	A_s	0.2 to 1.6	0.018
	Source thickness (m^2) [a]	δ	0.0025	0.0025
	Solid-phase diffusion coefficient (m^2/h) [b]	D	3.6×10^{-10}	3.6×10^{-10}
	Gas-phase mass transfer coefficient (m/h) [c]	h_m	1.44	5.04
	Material-air partition coefficient (-)	K	2.3×10^{11}	2.3×10^{11}
Sink	Area of chamber walls (m^2)	A_i	1.6	0.018
	Freundlich equilibrium constant [d]	K_f	3800	6000
	Freundlich index [d]	n	1.5	0.47

[a] Estimated by this author. [b] The original unit was m^2/s. [c] The original unit was m/s. [d] This parameter is defined by equation 14; as explained in Section 6.4.3, it is not used in the simulations.

Program i-SVOC includes two dynamic adsorption models for impermeable surfaces: the Langmuir and Freundlich adsorption models (See Sections 6.4.2 and 6.4.3 below for more details). However, this program does not support the adsorption models that are based on the assumption of instantaneous equilibrium. Thus, i-SVOC cannot use the sink parameters (K_f and n) listed in Table 6. The dynamic Freundlich adsorption model requires four parameters: f_a, α, f_d, and β (See equation 13 in Section 6.4.2). For the CLIMPAQ chamber, α was set to 1 and β 0.67. These values satisfy $n = \alpha/\beta = 1.5$ given in Table 6. Similarly, for the FLEC, α and β were set to 0.47 and 1.0, so their ratio is $n = 0.47$. In both cases, the non-linear adsorption and desorption rate constants (f_a and f_d) were estimated by trial and error. The estimated sink parameters are summarized in Table 7.

Table 7. Estimated parameters for the dynamic Freundlich adsorption model (see Section 6.4.3).

Parameter Name	Symbol	Value	
		CLIMPAQ	FLEC
Nonlinear adsorption rate constant (μg^{1-na} m^{3na-2} h^{-1})	f_a	5	10
Index for non-linear adsorption (dimensionless)	α	1	0.47
Nonlinear desorption rate constant (μg^{1-nd} m^{3nd-2} h^{-1})	f_d	0.025	0.0015
Index for non-linear desorption (dimensionless)	β	0.67	1

The method used to estimate parameters f_a and f_d in Table 7 was essentially a data-fitting process which, in this case, is time-consuming and requires some experience. A much simpler method for converting the equilibrium Freundlich model to a roughly equivalent dynamic Freundlich model is described in Appendix B.

3.6.2 Modeling the chamber concentrations

Two models are needed for the parameters listed in Tables 6 and 7, one for the CLIMPAQ and the other for the FLEC. To create a new model, click the <New> speed button (the first to the left) or select <File>, <New> from the main menu. This model will need the following pages/forms:

- <Building>
- <Sources> / <Diffusional sources>
- <Sinks> / <Langmuir & Freundlich sinks>
- <Conditions>

Now enter the parameters listed in Tables 6 and 7. In the <Conditions> page, enter the following information. Then, compile and inspect the model. Save the model to an external file.

Initial concentration in air:	0 ($\mu g/m^3$)
Simulation duration:	10000 (h)
Number of output data points:	200
Output options:	00) Air: gas-phase only ($\mu g/m^3$)

CAUTION: The model for the CLIMPAQ can be run at the normal speed, but the one for the FLEC cannot, as explained in Section 3.7. The latter should be run with the <Run slow> command, and it may take several minutes to complete the simulation because of the long duration.

As shown in Figures 22 and 23, the simulation results compare favorably to the experimental results (i.e., Figures 2 and 6 in Clausen et al., 2004).

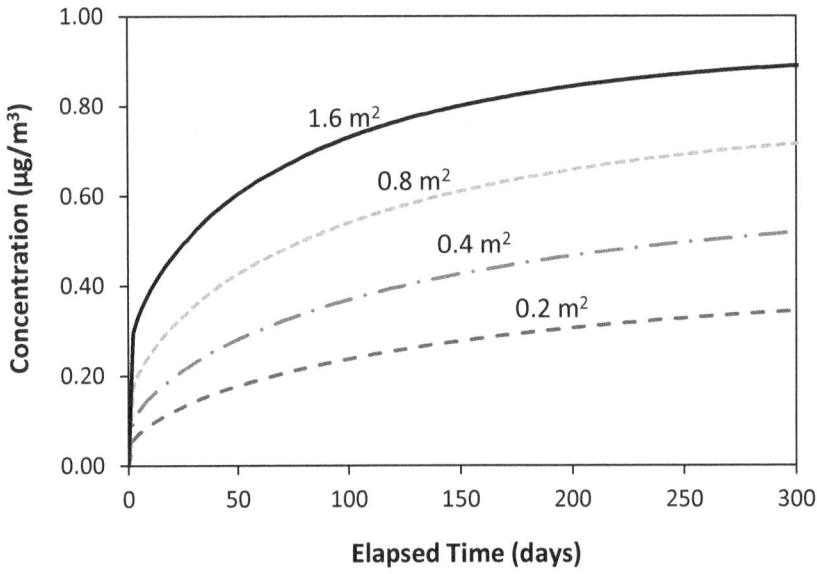

Figure 22. Simulated DEHP concentrations in the CLIMPAQ due to emissions from vinyl flooring (This figure was generated from four simulations with different source areas).

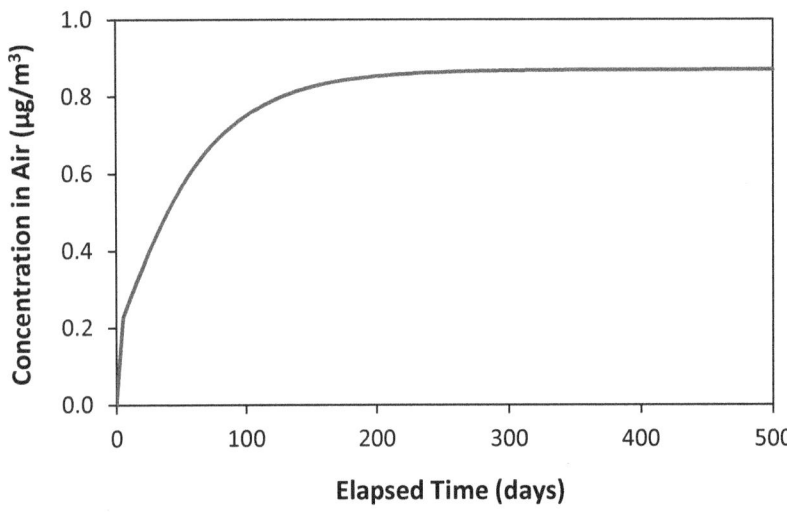

Figure 23. Simulated DEHP concentrations in the FLEC due to emissions from vinyl flooring.

3.7   When a simulation fails

3.7.1 Causes of simulation failures

Simulations may fail from time to time. Most failures are caused by the extreme values — in the context of numerical computation — in the input parameters. The following conditions are known to cause problems:

- The solid- or liquid-phase diffusion coefficient is very large (e.g., $> 10^{-8}$ m^2/h)
- The particle diameter is very small (e.g., <0.5 μm)
- The chamber volume is very small and the air flow rate very large (e.g., the FLEC).

Program i-SVOC provides the following remedies when a simulation cannot continue.

3.7.2 Using the <Run slow> command

When a simulation fails, the user can re-run his/her model at a slower pace by clicking the <Run slow> speed button (i.e. the one with the glyph of a walking man) or, equivalently, by selecting the <Run> / <Run slow> command from the main menu. This simulation mode can resolve many problems. For example, the <Run slow> command can handle particles as small as 0.001 μm in diameter. The tradeoff is that it takes more time to run a simulation.

3.7.3 Using the <Run very slow> command

The <Run very slow> command under the <Simulate> menu can handle even more difficult cases. However, this simulation mode is very time-consuming and, thus, should be used occasionally and only for short-term simulations (e.g., no longer than several hours).

3.7.4 A failed simulation will not hurt the operating system

When numerical integration fails, the program will halt the simulation and then exit gracefully without affecting the operating system. In such cases, an error message will be displayed.

4. More Features

4.1 Time-varying ventilation rates

Program i-SVOC allows simulations with time-varying ventilation rates. In the <Building> page, click the <Variable> radio button to display the <Ventilation rate changes> table, where the user defines all the ventilation rates other than the base ventilation rate. Demonstration model "Varying_ACH.svoc" uses the parameters shown in Table 8. To open a demonstration file, select <File> / <Open demo file> from the main menu. The <Building> page of the model is shown in Figure 24 and the simulation results in Figure 25.

Table 8. Example parameters for demonstration of time-varying ventilation rates.

Page/Form	Parameter Name	Value
Building	Room volume (m^3)	30
	Base ventilation rate (h^{-1})	1
	Ventilation rate between 24 and 48 hours	2
	Ventilation rate between 70 and 80 hours	3
Sources/ Other sources	Source type	Constant rate
	Emission rate ($\mu g/h$)	30
Conditions	Initial air concentration	0
	Simulation duration	100
	Output data points	100
	Output selection	00) Air: gas phase only ($\mu g/m^3$)

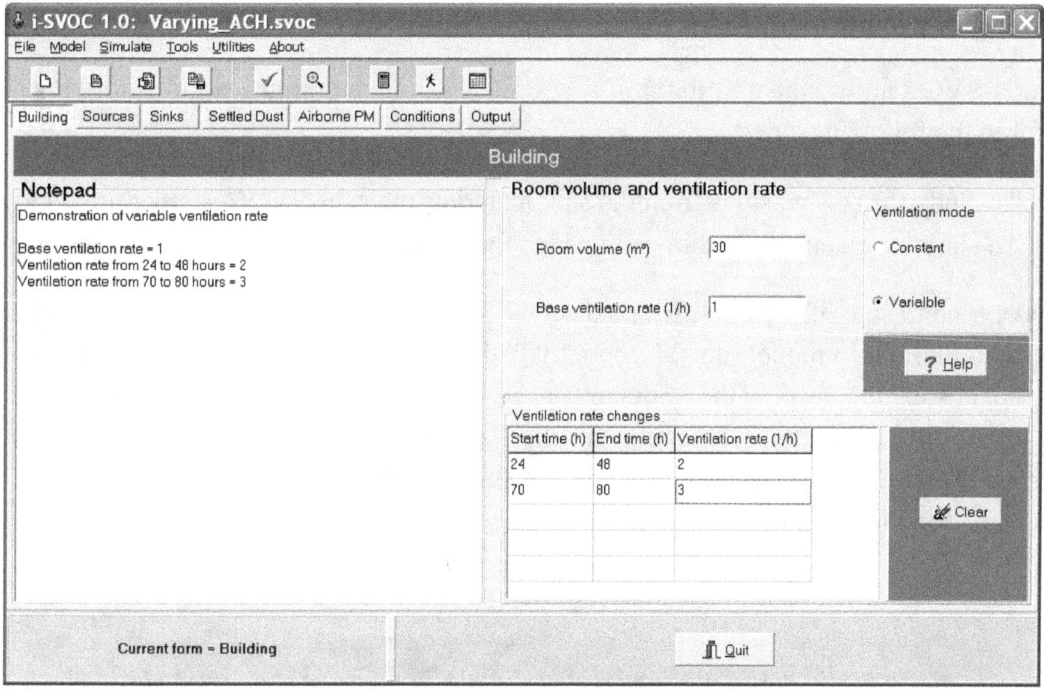

Figure 24. Defining variable ventilation rates (parameters from Table 8).

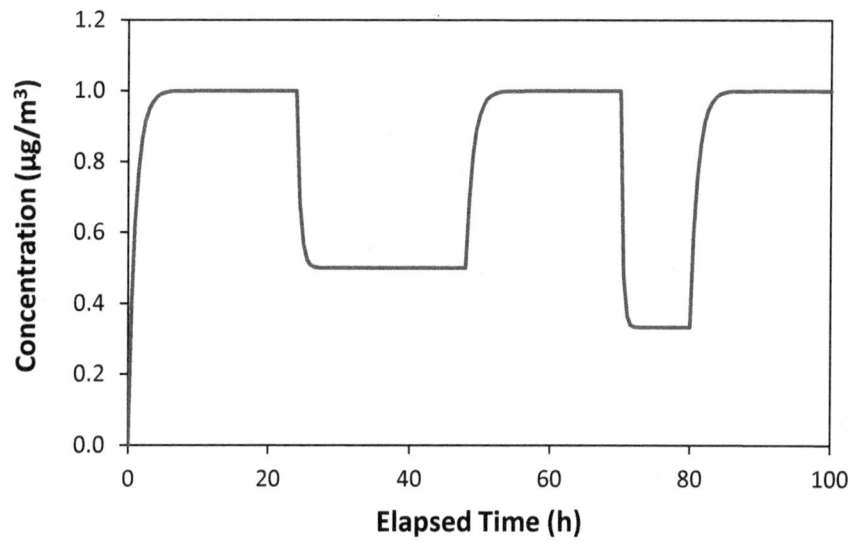

Figure 25. Simulation results for varying ventilation rates (file "Varying_ACH.svoc").

4.2   Source removal

Program i-SVOC allows the user to "stop" a source at any time during a simulation. This feature is useful in the following cases:

- To study the re-emissions from the sink materials after the SVOC sources are removed.
- To simulate chamber conditions in which the source is cut off during the experiment.

You have tried to shut off a non-diffusional source in Section 3.4.2. Shutting off a diffusional source is similar. Open model file "MyModel-02" that you created in Section 3.3. Go to the <Diffusional sources> form in the <Sources> page. To edit the source, click the source name and then click <Edit>. Check the source removal check box and then enter 7000 for the "removal time" (see Figure 26). Now re-run the model. As shown in Figure 27, the re-emission from the sink can last for a long time.

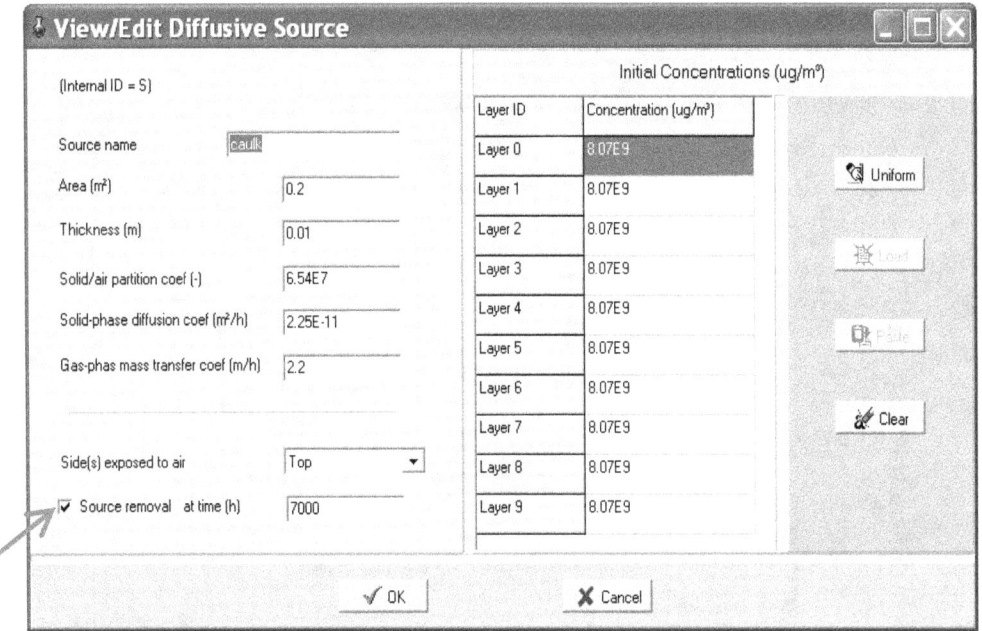

Figure 26. Defining the source removal time.

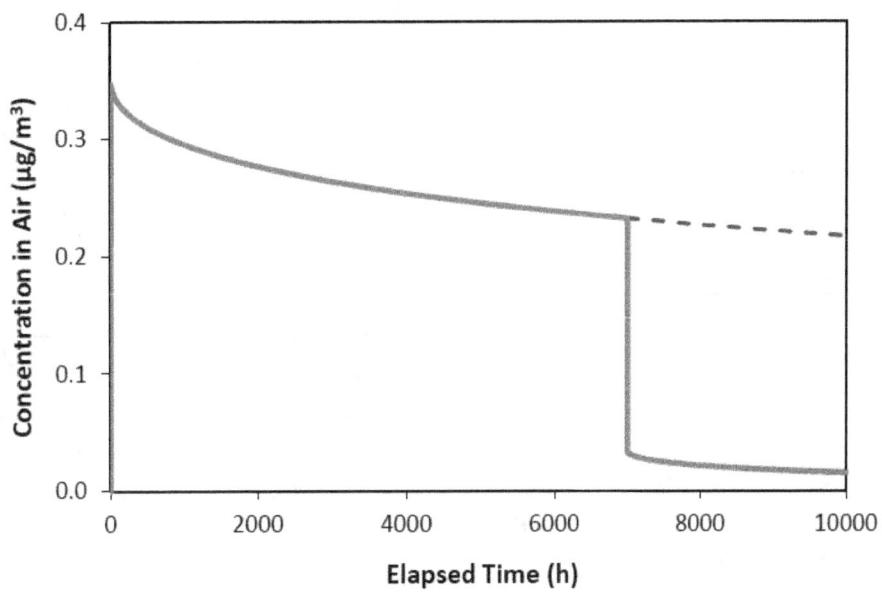

Figure 27. Air concentrations before and after the source removal at 7000 elapsed hours.

4.3 Double-layer sources

Program i-SVOC can simulate double-layer sources (e.g., painted wood board and laminated composite wood board). Two steps are needed to create a double-layer source:

- Define the top and bottom layers as two single-layer sources, and
- Use the <Group> command to combine the two.

For demonstration purposes, we will create a model with the parameters listed in Table 9.

Table 9. Example parameters for simulating a double-layer source.

Page/Form	Parameter Name	Values	
Building	Room volume (m^3)	30	
	Ventilation rate (h^{-1})	1	
Sources/ Diffusional sources	Source area (m^2)	1	
	Gas-phase mass transfer coefficient	1	
	Layer ID	Top	Bottom
	Material thickness (m)	0.001	0.01
	Solid-air partition coefficient (dimensionless)	1.0×10^6	2.0×10^7
	Solid-phase diffusion coefficient (m^2/h)	1.0×10^{-10}	5.0×10^{-11}
	Initial concentration in layer (µg/m^3)	2.0×10^7	2.0×10^9
	Side(s) exposed	Top	Neither
Simulation conditions	Initial concentration in air (µg/m^3)	0	
	Simulation duration (h)	10000	
	Number of output data points	200	
	Output options	00) Air: gas-phase only (µg/m^3)	

After entering two single-layer sources, the <Diffusional sources> form should look like Figure 28. Note that, although the bottom layer does not need the gas-phase mass transfer coefficient, the program still requires a numerical value, which eventually will be ignored.

Now click the <Group> button to bring up the source grouping form (Figure 29). On the left side of the form, select the top and bottom layers. Ignore the right side, which is for encapsulation. Click <OK> to return to the < Diffusional sources> form. Note the changes in the top row (Figure 30). Run the model. Figure 31 was created by demonstration file "2-layer-1.svoc."

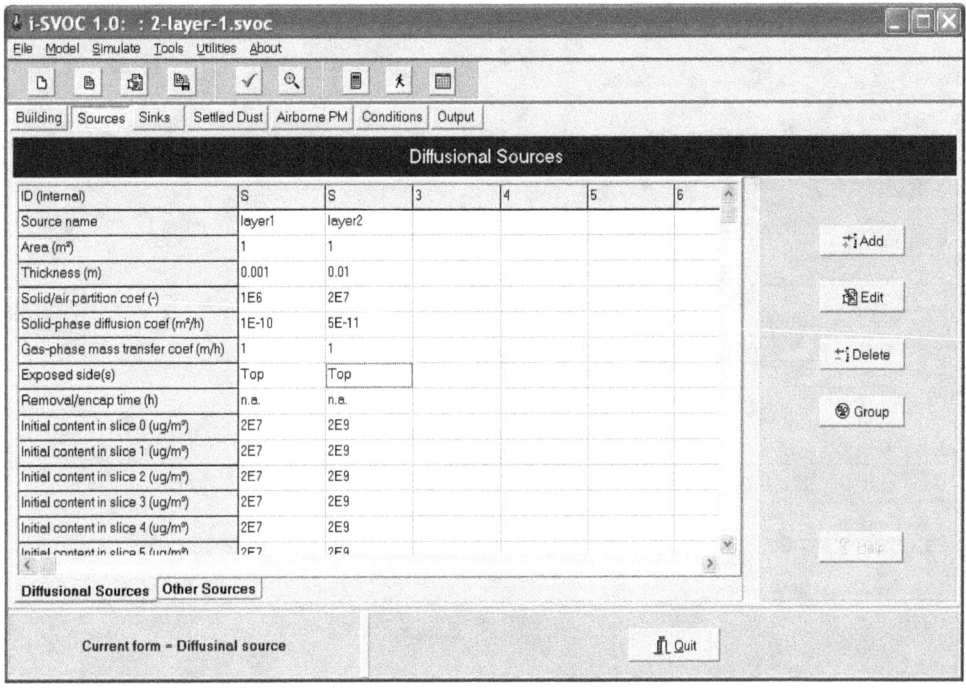

Figure 28. The double-layer source was first created as two single-layer sources.

Creating a double-layer source or sink

Is the top layer an encapsulant?

Select the top layer

layer1

Select the bottom layer

layer2

(•) No   ( ) Yes

Enter time for encapsulation (h)

Edit1

√ OK   ✗ Cancel

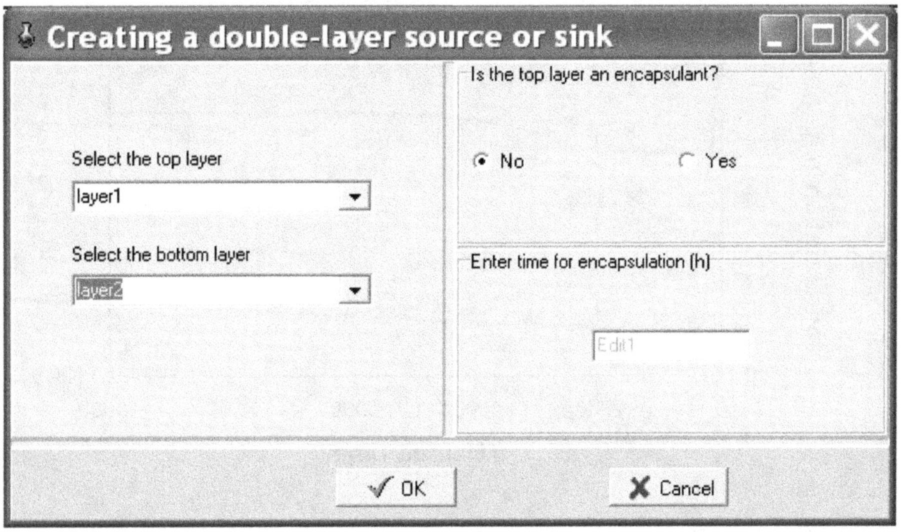

Figure 29. The form for grouping two single-layer sources into a double-layer source.

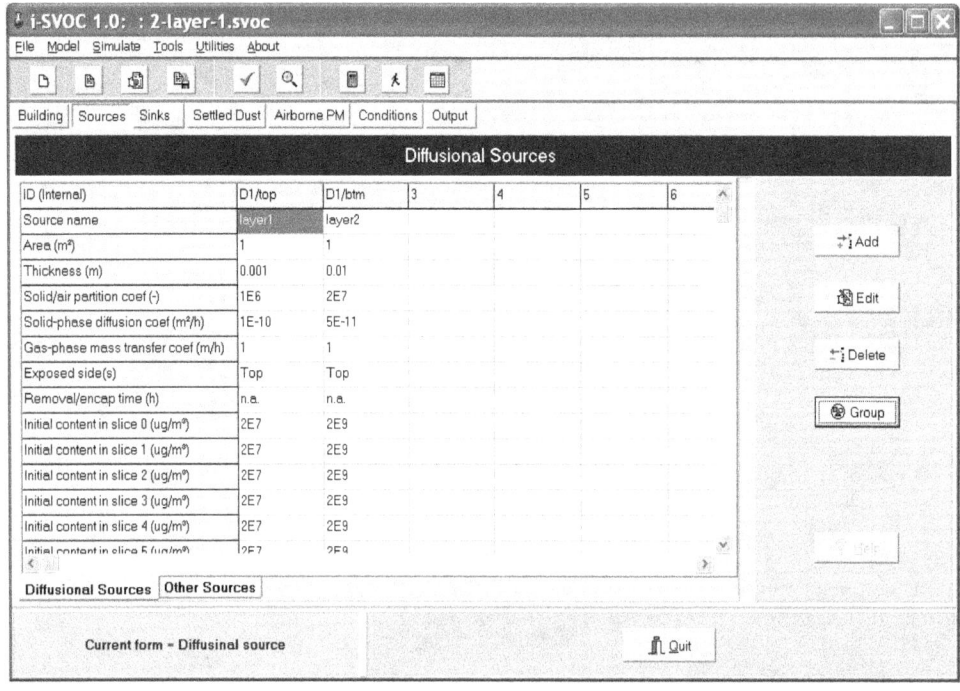

Figure 30. The two single-layer sources have been grouped into a double-layer source. Note the changes in source IDs.

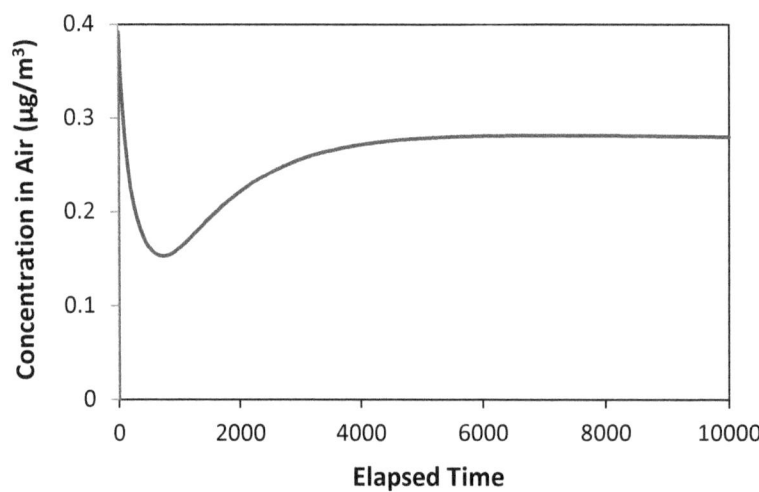

Figure 31. Air concentration due to emissions from a hypothetical double-layer source (see Table 9).

4.4 Encapsulation

An encapsulated source is equivalent to a double-layer source except that the initial concentration in the encapsulant layer is often zero. Parameters used by demonstration program "encap-1.svoc" are given in Table 10.

Table 10. Example parameters for simulating an encapsulated source.

Page/Form	Parameter Name	Value	
Building	Room volume (m^3)	30	
	Ventilation rate (h^{-1})	1	
Source/ Diffusional Sources	Source area (m^2)	1	
	Gas-phase mass transfer coefficient	1	
	Layer ID	Encapsulant (top)	Source (Bottom)
	Material thickness (m)	0.001	0.01
	Solid-air partition coefficient (dimensionless)	1.0×10^6	2.0×10^7
	Solid-phase diffusion coefficient (m^2/h)	1.0×10^{-10}	5.0×10^{-11}
	Initial concentration in layer (μg/m^3)	2.0×10^7	2.0×10^9
	Side(s) exposed	Top	Top [a]
	Time for encapsulation (h)	2000	n.a.
Simulation conditions	Initial concentration in air (μg/m^3)	0	
	Simulation duration (h)	10000	
	Number of output data points	200	
	Output options	00) Air: gas-phase only (μg/m^3)	

[a] Before encapsulation, the top-side of the source is exposed to air.

Creating an encapsulated source is similar to creating a double-layer source. During the "grouping" process, the user can specify the time when the encapsulant is applied (Figure 32).

The next step is to edit the parameters for the bottom layer by setting the "Exposed side(s)" to "Top." Figure 33 was created by demonstration file "encap-1.svoc."

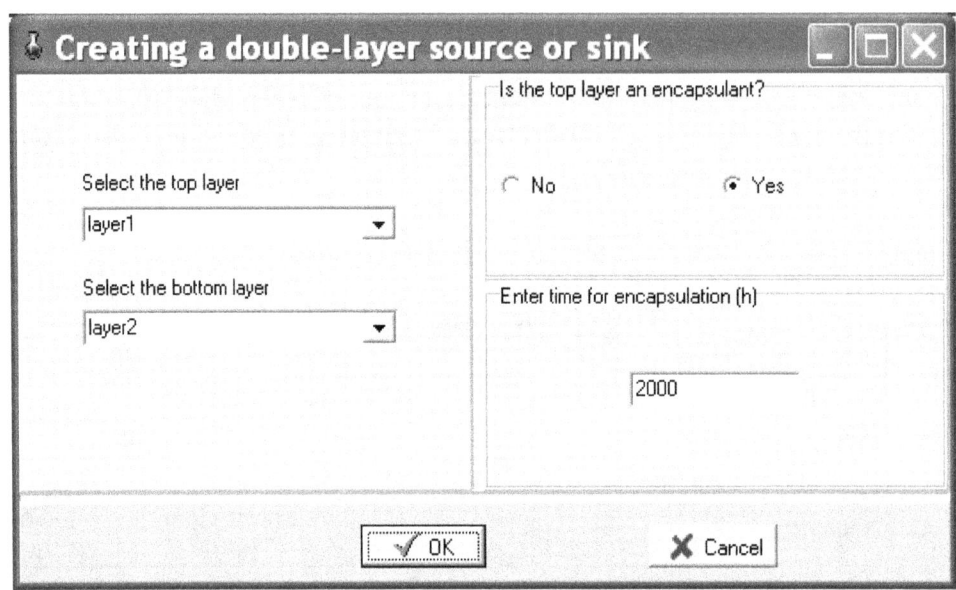

Figure 32. Define the encapsulation time in the source-grouping form.

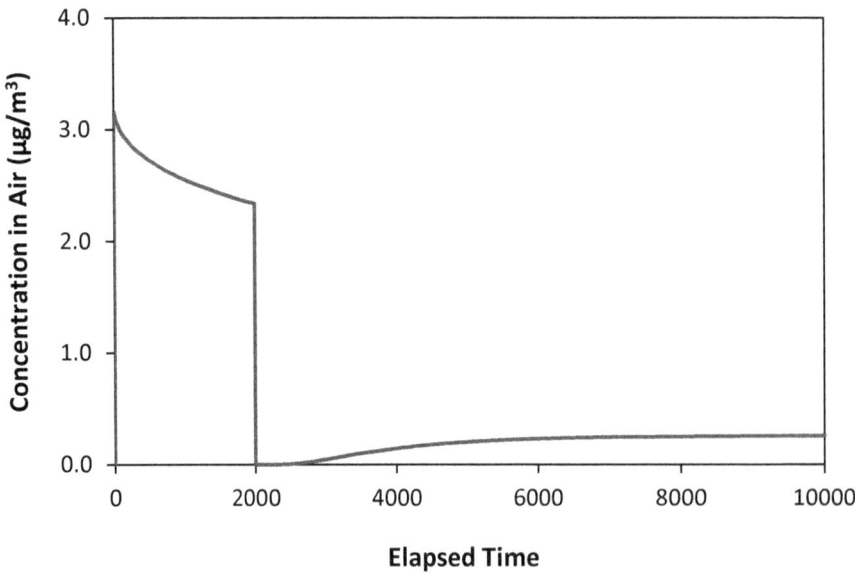

Figure 33. Simulated effect of source encapsulation on indoor air concentration using hypothetical parameters in Table 10.

44

4.5   Pulse release of airborne particles

When particles are injected into room air in a pulse, all the particles have the same age and thus the same residence time. These conditions make the simulation results much easier to interpret than those for other particle release modes.

Figure 34 was created with demonstration model "Airborne PM-pulse.svoc", in which 1μm-diameter particles are released into room air at 1, 2, and 3 elapsed hours. Please open the model file for more details. Note that, in the pulse release mode, the PM number concentration is the initial concentration for the pulse release, and that you can create multiple pulse releases for the same particles and for different particles.

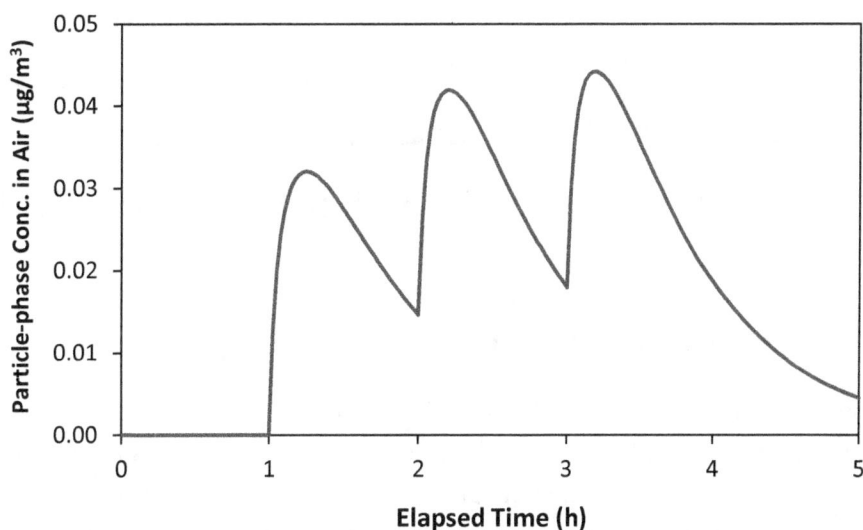

Figure 34. Simulated total particle-phase SVOC concentration in air due to three consecutive pulse releases of particles; created by demonstration model "Airborne PM-pulse.svoc".

4.6   Episodic sources for airborne particles

Many indoor PM sources such as particle resuspension, cooking, smoking, and incense burning, are episodic in nature. These sources can be represented by a series of pulse releases. Program i-SVOC includes a calculation sheet to assist the user to convert the expected number concentration during the episode to a series of pulse releases. To bring up the calculation sheet

(Figure 35), select <Simulating episodic PM sources> from the <Tools> menu. The calculation method is described in Section 6.8.3.

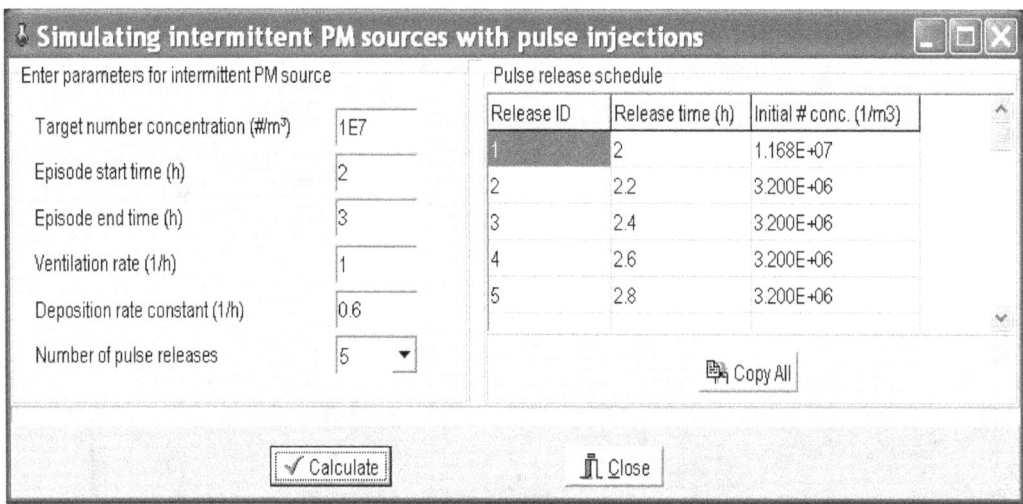

Figure 35. Calculation sheet for simulating an episodic source by a series of pulse releases.

4.7   Using the <Run and save> command

Some legacy pollutants may have existed in a building for decades and a pseudo-steady state is often reached. In these cases, the concentration gradients within the diffusional sources and sinks are unique to the specific building. Modelers are often interested in the current status instead of the history of the contamination. The <Run and save> command allows the user to simulate the contamination history for a given building and save the current status for future use. When you run a model with the <Run and save> command, the results at the end of the simulation will be put back to the model. For demonstration, open "MyModel-02.svoc" and then select <Simulate>, <Run and save> from the main menu. At the end of the simulation, the concentration gradients will be posted (Figure 36). If you save it with a new file name, it can be used as the starting point for new simulations. This feature works only for diffusional sources and sinks. Also note that the SVOC concentration in room air at the end of the simulation will also be posted to the <Conditions> page.

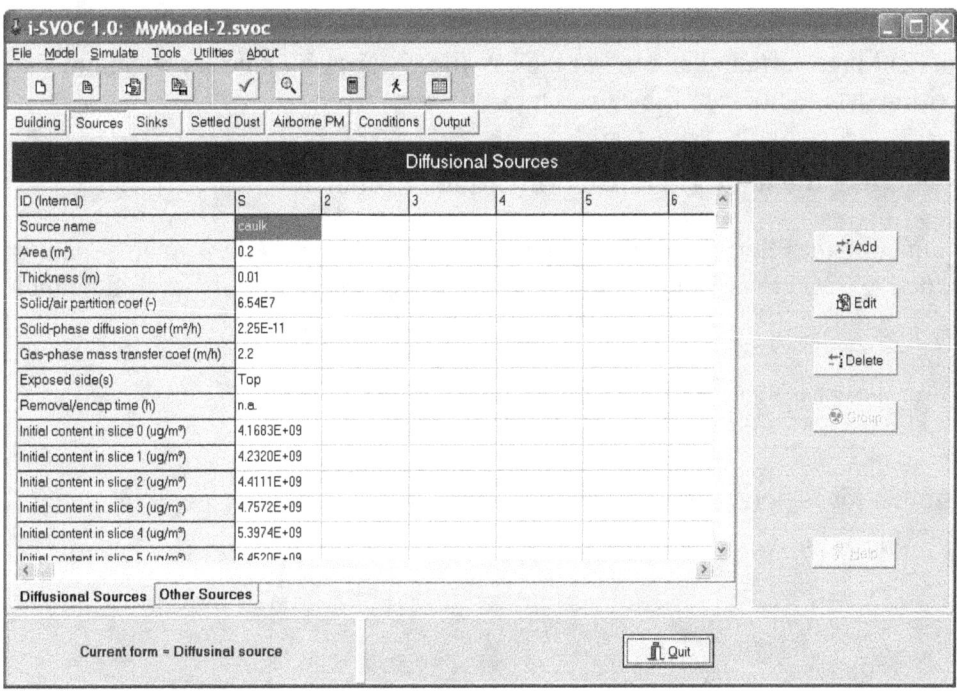

Figure 36. After running model "MyModel-02" with the <Run and save> command, the concentration gradient in the source at 10000 hours is posted on the source form. Similar changes were made to the sink form.

4.8 Using the <Run batch> command

The <Run batch> command under the <Simulate> menu allows the user to run multiple models unattended. This feature is especially useful for screening multiple chemicals and for models that take a long time to run.

To use this feature, all the models should be saved in the same folder. It is highly recommended but not required to compile each model. After clicking on <Simulate> / <Run batch>, an open-file dialog will appear. Select model files and then click the <Open> button. The program will run the selected models sequentially. A run-batch report is available after the work is done.

The output data will be saved in the same folder where the model files are located. The output file names will be the same as the model files except with the "csv" extension, which stands for comma separated values. Below is an example:

Model file	MyModel-xyz.svoc
Output for concentrations	MyModel-xyz.csv
Output for mass fluxes	MyModel-xyz(Rate).csv

CAUTIONS: (1) If you would like to re-run a batch for some reason, the old csv files will be overridden. (2) Make certain that none of the csv files is open by other applications such as Excel because writing data to an open file causes an I/O error.

4.9   Calculating the total SVOC concentration in room air

This program computes the gas-phase and particle-phase concentrations in air separately. To obtain the total SVOC concentration in air, the following output options should be included in the model.

00) Air: gas-phase only ($\mu g/m^3$)
09) Airborne PM: total particle-phase SVOC ($\mu g/m^3$ air)

After the simulation is done, copy and paste the results to a spreadsheet and then sum the two data columns up.

5. Program Specifications

The program specifications are provided in Tables 11 and 12. In Table 11, the total number of ordinary differential equations is an overall limiting factor. When a model exceeds the maximum allowable number of differential equations, the user will receive an error message.

Table 11. Maximum allowable number of components in a model.

Model Components	Max.	Notes
SVOC species	1	
Air zone	1	
Ventilation rate changes	6	[a]
Single-layer diffusional sources or sinks	10	
Double-layer diffusional sources or sink	3	
Non-diffusional sources	10	
Langmuir/Freundlich sinks	10	
Airborne particle type I (permeable)	10	[b]
Airborne particle type II (surface sorption)	10	[b]
Airborne particle type III (liquid film over solid core)	10	[b]
Settled particle type I (permeable)	10	[b]
Settled particle type II (surface sorption)	10	[b]
Total number of ordinary differential equations	150	[c]
Maximum simulation duration (h)	5×10^6	
Maximum number of output data points	1000	

[a] One base ventilation rate plus 5 changes.

[b] Number of size bins or particle types (i.e., same size but different properties).

[c] When the number of equations exceeds this limit, an error message will be displayed.

Table 12. List of options for simulation output.

Data Group	Output Option	Notes
	00) Air: gas-phase only ($\mu g/m^3$ air)	Gas-phase SVOC concentration in room air
	01) Diffusional sources/sinks: individual slices ($\mu g/m^3$ solid)	SVOC concentrations as a function of depth in the material
	02) Diffusional sources/sinks: average ($\mu g/m^3$ solid)	Average SVOC concentrations in the material
	03) Surface adsorption: Langmuir or Freundlich sinks ($\mu g/m^2$ surface)	For non-diffusional sinks
SVOC concentrations	04) Settled dust: size-segregated concentrations in dust ($\mu g/g$ dust)	SVOC concentrations in settled dust by size bins
	05) Settled dust: average concentration in dust ($\mu g/g$ dust)	Weighted mean concentration in dust, from Eq. 51
	06) Settled dust: individual hollow spheres (Type I only) ($\mu g/m^3$ dust)	SVOC concentrations in settled dust as a function of depth inside the particle and by size bins
	07) Airborne PM: size-segregated particle-phase SVOC ($\mu g/g$ PM)	Size-segregated particle-phase SVOC concentrations in room air
	08) Airborne PM: average particle-phase SVOC ($\mu g/g$ PM)	Weighted mean of all PM sizes and types, from Eq. 47
	09) Airborne PM: total particle-phase SVOC ($\mu g/m^3$ air)	Weighted sum of all PM sizes/types, from Eq. 49
Airborne PM concentrations	10) Airborne PM: total number concentration (counts/m^3 air)	Total number concentration of airborne particles
	11) Airborne PM: total mass concentration (μg PM/m^3 air)	Total mass concentration of airborne particles
SVOC fluxes	12) SVOC mass fluxes ($\mu g/m^2/h$)	Positive values mean emissions; negative values mean sorption

6. Inside i-SVOC

6.1 Programming language and supporting software

Program i-SVOC was written in Delphi XE Professional Edition (Embarcadero Technologies, San Francisco, CA). The installation package was created by InstallSimple 2.7 PRO (InstallSimple Solutions, www.installsimple.com).

6.2 Numerical method

The ordinary differential equations are solved by the fourth/fifth-order Runge-Kutta-Fehlberg method (Cheney and Kincaid, 1980). The original code was written in Fortran (Forthythe et al., 1977). This author translated it into Delphi and made several minor adjustments to the code.

6.3 Modified state-space method

In program i-SVOC, the diffusion processes inside a solid material or particle are represented by the modified state-space (MSS) method, which divides the solid phase into a finite number of slices or hollow spheres. The methods for dividing the flat materials and particles are presented below. More details about the method development and validation can be found in Guo (2013 and 2014).

6.3.1 Diffusional sources and sinks

The method for dividing a flat material is as follows: (1) The thickness of the exposed slice is ultra thin (1×10^{-6} m); (2) The inner slices are divided according to equation 1:

$$\Delta L_j = 2^p \, \Delta L_i \qquad (1)$$

where ΔL_j = thickness of the inner slice of two adjacent slices (m)
ΔL_i = thickness of the outer slice of two adjacent slices (m)
p = 0 or 1.

The value of p is determined by the thickness and solid-phase diffusion coefficient of the substrate, as shown in Table 13.

Table 13. Selection of the slicing methods for diffusional sources and sinks.

Substrate thickness (L) (mm)	Value of D_m (m²/h)	Value of p for Inner slices	Number of slices
$L < 2$	n/a	0	10
$L \geq 2$	$>10^{-9}$	0	10
	$\leq 10^{-9}$	1	10

6.3.2 Permeable particles

The method for dividing spherical particles is similar except that the flat "slices" become "hollow spheres" (Guo, 2014). The thickness of the top hollow sphere is set to 1×10^{-7} m (i.e., 0.1 μm) unless the particle diameter is equal to or less than 0.25 μm (Table 14).

Table 14. Selection of the slicing method for permeable particles.

Particle diameter (d) (μm)	Number of layers [a] (n)	Thickness of top hollow sphere (μm)	Value of p for inner hollow spheres
$d \leq 0.25$	1	$d/2$ [b]	n/a
$0.25 < d \leq 1$	2	0.1	n/a
$1 < d \leq 5$	3	0.1	1
$d > 5$	5	0.1	1

[a] Including (n-1) hollow spheres and one solid sphere. [b] This is a solid sphere.

6.4   Mass transfer equations

This section contains all the mass transfer equations used in the program. More detailed can be found in Guo (2013, 2014) and the references herein.

6.4.1 Diffusional sources and sinks

Mass transfer between the top slice and room air:

$$R_{ma} = A H_a \left(\frac{C_{m1}}{K_{ma}} - C_a \right) \qquad (2)$$

52

$$\frac{1}{H_a} = \frac{1}{K_{ma}\, h_m} + \frac{1}{h_a} \tag{3}$$

$$h_m = \frac{2\, D_m}{\Delta L_1} \tag{4}$$

where R_{ma} = rate of mass transfer from the top (exposed) slice to air (μg/h)
  A = exposed area of the source or sink (m^2)
  H_a = overall gas-phase mass transfer coefficient (m/h) from equation 3 (m/h)
  C_{m1} = SVOC concentration in the top (exposed) slice of the source or sink (μg/m^3)
  K_{ma} = solid-air partition coefficient (dimensionless)
  C_a = SVOC concentration in room air (μg/m^3).
  h_a = gas-phase mass transfer coefficient (m/h)
  h_m = solid-phase mass transfer coefficient, from equation 4 (m/h)
  D_m = solid-phase diffusion coefficient (m^2/h)
  ΔL_1 = thickness of the top (exposed) slice (m).

Note that the solid-phase mass transfer coefficient is not only a function of diffusion coefficient but also a function of the thickness of the slice.

Mass transfer between two adjacent slices of the same material:

$$R_{ij} = A\, h_m \left(C_{mi} - C_{mj} \right) \tag{5}$$

$$h_m = \frac{2\, D_m}{\Delta L_i + \Delta L_j} \tag{6}$$

where R_{ij} = rate of mass transfer from slice i to slice j (μg/h)
  h_m = solid-phase mass transfer coefficient (m/h)
  D_m = solid-phase diffusion coefficient (m^2/h)
  ΔL_i, ΔL_j = thicknesses of slices i and j (m)
  $(\Delta L_i + \Delta L_j)/2$ = travel distance for inter-slice diffusion (m)
  C_{mi} = concentration in slice i (μg/m^3)
  C_{mj} = concentration in slice j (μg/m^3).

Mass transfer between two adjacent slices of the different materials:

This type of mass transfer applies only to double-layer sources or sinks. Equation 7 is for the mass transfer across the material-material interface:

53

$$R_{21} = A H_{m1} \left(K_{12} C_{m2} - C_{m1} \right) \qquad (7)$$

$$\frac{1}{H_{m1}} = \frac{1}{h_{m1}} + \frac{K_{12}}{h_{m2}} \qquad (8)$$

$$h_{m1} = \frac{2 D_{m1}}{\Delta L_1} \qquad (9)$$

$$h_{m2} = \frac{2 D_{m2}}{\Delta L_2} \qquad (10)$$

$$K_{12} = \frac{K_{ma1}}{K_{ma2}} \qquad (11)$$

where R_{21} = rate of mass transfer from material 2 to material 1 (μg/h)
A = area of the material/material interface (m^2)
H_{m1} = overall mass transfer coefficient with respect to material 1, from equation 8 (m/h)
K_{12} = solid-solid partition coefficient for materials 1 and 2, from equation 11 (dimensionless)
C_{m2} = concentration in the slice of material 2 in contact with material 1 (μg/m^3)
C_{m1} = concentration in the slice of material 1 in contact with material 2 (μg/m^3)
h_{m1}, h_{m2} = solid-phase mass transfer coefficients for materials 1 and 2 (m/h)
D_{m1}, D_{m2} = solid-phase diffusion coefficients for materials 1 and 2 (m^2/h)
ΔL_1, ΔL_2 = thicknesses of the two contacting slices for materials 1 and 2 (m)
K_{ma1}, K_{ma2} = material/air partition coefficients for materials 1 and 2 (dimensionless).

6.4.2 Dynamic Langmuir sink

The net adsorption rate for the dynamic Langmuir sink is given by equation 12 (Tichenor, et al., 1991):

$$R_L = A \left(k_a C_a - k_d C_s \right) \qquad (12)$$

54

where R_L = net adsorption rate for the Langmuir sink (μg/h)
  A = area of sink surface (m^2)
  k_a = adsorption rate constant (m/h)
  C_a = concentration in air (μg/m^3)
  k_d = desorption rate constant (h^{-1})
  C_s = concentration on sink surface (μg/m^2).

6.4.3 Dynamic Freundlich sink

The net adsorption rate for the dynamic Freundlich sink is given by equation 13 (Van Loyd et al., 1997):

$$R_F = A\left(f_a\, C_a^{\,\alpha} - f_d\, C_s^{\,\beta} \right) \tag{13}$$

where R_F = net adsorption rate for the Freundlich sink (μg/h)
  A = area of sink surface (m^2)
  f_a = nonlinear adsorption rate constant (μg$^{1-\alpha}$ m$^{3\alpha-2}$ h^{-1})
  f_d = nonlinear desorption rate constant (μg$^{1-\beta}$ m$^{2\beta-2}$ h^{-1})
  C_a = concentration in air (μg/m^3)
  C_s = concentration on sink surface (μg/m^2)
  α and β = dimensionless constants.

Note that, at steady state, equation 13 yields the Freundlich adsorption isotherm (equation 14):

$$C_s = K_f\, C_a^{\,n} \tag{14}$$

where K_f and n are constants for a given chemical and sink material at a given temperature, calculated from equations 15 and 16.

$$K_f = \left(\frac{f_a}{f_d} \right)^{1/\beta} \tag{15}$$

$$n = \frac{\alpha}{\beta} \tag{16}$$

This program does not support the equilibrium model (equation 14). However, it can be converted to a roughly equivalent dynamic model (equation 13) if parameters K_f and n are known. Details are given in Appendix B.

CAUTION: The behavior of the Freundlich adsorption models (equations 13 and 14) is poorly understood. Little is known about the dependence of the model parameters on the properties of the adsorbate and adsorbent. Thus, they should be used with caution.

6.4.4 Permeable particles

SVOC interactions with permeable particles, either suspended or settled, are represented by the modified state-space (MSS) method, which divides the particle of a given size into several hollow spheres and one solid sphere at the core (Guo, 2014).

Mass transfer between room air and the top hollow sphere:

$$R_{ap} = A H_a \left(C_a - \frac{C_{pl}}{K_{ma}} \right)$$

(17)

$$\frac{1}{H_a} = \frac{1}{K_{pa} h_p} + \frac{1}{h_a}$$

(18)

$$h_p = \frac{D_p}{(r_0 - r_{td})}$$

(19)

where R_{ap} = rate of mass transfer from room air to airborne particles (µg/h)
     A = surface area of the particle (m^2)
     H_a = overall gas-phase mass transfer coefficient, from equation 18 (m/h)
     C_{pl} = concentration in the exposed hollow sphere (µg/m^3)
     K_{pa} = particle-air partition coefficient (dimensionless)
     C_a = concentration in room air (µg/m^3).
     h_p = particle-phase mass transfer coefficient, from equation 19 (m/h)
     D_p = particle-phase diffusion coefficient (m^2/h)
     r_0 = radius of the particle (m)
     r_{td} = radius that divides the top hollow sphere into two parts with equal volumes (equation 20).

$$\frac{4}{3}\pi\,(r_0^3 - r_{td}^3) = \frac{4}{3}\pi\,(r_{td}^3 - r_1^3) \tag{20}$$

where r_0 = outside radius of the top hollow sphere (m)
   r_1 = inside radius of the top hollow sphere (m).

Mass transfer within the particle:

The rate of mass transfer between two adjacent hollow spheres, i and j, is determined by equation 21:

$$R_{ij} = A_i\,h_p\left(C_{pi} - C_{pj}\right) \tag{21}$$

$$A_i = 4\pi r_i^2 \tag{22}$$

$$h_p = \frac{D_p}{r_{tdi} - r_{tdj}} \tag{23}$$

where R_{ij} = rate of mass transfer from hollow sphere i to hollow sphere j (μg/h)
   A_i = contact area for hollow spheres i and j, from equation 22 (m^2)
   h_p = particle-phase mass transfer coefficient between hollow spheres i and j, from equation 23 (m/h)
   r_i = inside radius of hollow sphere i (i.e., the outer hollow sphere) (m)
   r_{tdi} = radius for travel distance in hollow sphere i, from equation 20 (m)
   r_{tdj} = radius for travel distance in hollow sphere j, from equation 20 (m)
   $(r_{tdi} - r_{tdj})$ = travel distance for diffusion between the two hollow spheres (m).

6.4.5 Impermeable particles

The surface adsorption of SVOCs by impermeable particles is represented by the dynamic forms of the Langmuir or Freundlich adsorption models (equations 12 and 13).

6.4.6 Particles with a liquid film over an impermeable solid core

The rate of mass transfer between indoor air and the liquid film is determined by equation 24:

$$R_{ap} = A_p\,H_a\left(C_a - \frac{C_L}{K_{La}}\right) \tag{24}$$

$$\frac{1}{H_a} = \frac{1}{K_{La}\, h_L} + \frac{1}{h_a} \qquad (25)$$

$$h_L = \frac{D_L}{r_0 - r_{td}} \qquad (26)$$

where R_{ap} = rate of mass transfer from indoor air to the particle (μg/h)
   A_p = surface area of the particle (m^2)
   H_a = overall gas-phase mass transfer coefficient, from equation 25 (m/h)
   C_a = gas-phase concentration in indoor air (μg/m^3)
   C_L = concentration in the liquid film of the particle (μg/m^3)
   K_{La} = liquid-air partition coefficient (dimensionless)
   h_L = liquid-phase mass transfer coefficient, from equation 26 (m/h)
   h_a = gas-phase mass transfer coefficient (m/h)
   D_L = liquid-phase diffusion coefficient (m^2/h)
   r_0 = radius of the particle (m)
   r_{td} = radius for travel distance in the liquid film, from equation 20 (m)
   $r_0 - r_{td}$ = travel distance in the liquid film (m).

6.4.7 SVOC mass fluxes

The SVOC mass fluxes are calculated from equation 27, where the mass transfer rate (R) is from equations 2, 12, 13, 17, or 24.

$$X = \frac{R}{A} \qquad (27)$$

where X = SVOC mass flux (μg/m^2/h)
   R = mass transfer rate (μg/h)
   A = contact area of the two phases (m^2).

The mass flux is a signed value. In this program, a positive flux means SVOC emission into room air while a negative value means sorption from room air.

6.5   Differential equations

All i-SVOC models are in the form of a system of first-order, ordinary differential equations. The differential equations for individual indoor media (i.e., compartments) are presented in Sections 6.5.1 through 6.5.7 below.

6.5.1 Room Air

The mass balance for the gas-phase concentration is given by equation 28:

$$V\frac{dC_a}{dt} = Q(pC_{amb} - C_a) + \sum_{i=1}^{n_1} R_{sai} - \sum_{j=1}^{n_2} R_{anj} - \sum_{k=1}^{n_3} n_{Dk} R_{adk} - V\sum_{l=1}^{n_4} N_{pl} R_{apl} \quad (28)$$

where V = room volume (m^3)

C_a = gas-phase SVOC concentration (μg/m^3)

t = time (h)

Q = air change flow rate (m^3/h)

p = penetration factor for the gas-phase SVOC in ambient air (fraction)

C_{amb} = gas-phase SVOC concentration in ambient air (μg/m^3)

R_{sai} = rate of emission from source i (μg/h)

R_{anj} = rate of mass transfer from gas phase to sink material j (μg/h)

n_{Dk} = total number of settled dust particles in size bin/particle type k in the room (counts)

R_{adk} = rate of mass transfer from gas phase to a single settled dust in size bin/dust type k (μg/h)

N_{pl} = number concentration of airborne particles in size bin/particle type l (counts/m^3)

R_{apl} = rate of mass transfer from gas phase to a single airborne particle in size bin/particle type l (μg/h)

n_1 = number of sources

n_2 = number of sinks

n_3 = number of size bins/particle types for settled dust

n_4 = number of size bins/particle types for airborne particles.

6.5.2 Diffusional sources and sinks

Equations 29, 30, and 31 are for, respectively, the exposed slice, interior slices, and bottom slice with no mass transfer:

For the top slice that is exposed to air:

$$v_1\frac{dC_{m1}}{dt} = -R_{1a} + R_{21} \quad (29)$$

For interior slices within the same material:

$$v_j\frac{dC_{mj}}{dt} = -R_{ji} + R_{jk} \quad (30)$$

59

For the bottom slice (n) without mass transfer to other media:

$$v_n \frac{dC_{mn}}{dt} = -R_{nn-1} \tag{31}$$

where v_1, v_j, v_n, = volumes of top slice, interior slice j, and bottom slice (m³)
C_{m1}, C_{mj}, C_{mn}, = concentrations in top slice, interior slice j, and bottom slice (μg/m³)
R_{1a} = rate of mass transfer from the top slice to air, from equation 2 (μg/h)
R_{21} = rate of mass transfer from the slice 2 to slice 1, from equation 5 (μg/h)
R_{ji} = rate of mass transfer from the slice j to slice i ($j=i+1$), from equation 5 (μg/h)
R_{jk} = rate of mass transfer from the slice j to slice k ($k=j+1$), from equation 5 (μg/h)
R_{nn-1} = rate of mass transfer from the slice n (the bottom slice) to slice n-1, from equation 5 (μg/h).

6.5.3 Surface adsorption

Adsorption of SVOCs on impermeable surface materials is represented by the dynamic form of the Langmuir model (equation 32) or the Freundlich model (equation 33):

$$A \frac{dC_s}{dt} = R_L \tag{32}$$

$$A \frac{dC_s}{dt} = R_F \tag{33}$$

where A = surface area (m²)
C_s = concentration on sink surface (μg/m²)
t = time (h)
R_L = net adsorption rate for dynamic Langmuir model, from equation 12 (μg/h)
R_F = net adsorption rate for dynamic Freundlich model, from equation 13 (μg/h).

6.5.4 Permeable particles

For permeable particles, equations 34 and 35 are for, respectively, the top and interior hollow spheres; equation 36 is for the solid sphere at the core.

$$v_1 \frac{dC_{p1}}{dt} = R_{ap} - R_{12} \qquad (34)$$

$$v_j \frac{dC_{pj}}{dt} = R_{ij} - R_{jk} \qquad (35)$$

$$v_n \frac{dC_{pn}}{dt} = -R_{nn-1} \qquad (36)$$

where v_1, v_j, v_n = volumes of hollow spheres 1 and j, and solid sphere n, from equation 37 (m^3)
C_{p1}, C_{pj}, C_{pn} = SVOC concentrations in hollow spheres 1 and j, and solid sphere n (μg/m^3)
R_{ap} = rate of mass transfer from air to the top hollow sphere, from equation 17 (μg/h)
R_{12} = rate of mass transfer from the top hollow sphere to the adjacent hollow sphere, from equation 21 (μg/h)
R_{ij} = rate of mass transfer from hollow sphere i to hollow sphere j (j=i+1), from equation 21 (μg/h)
R_{jk} = rate of mass transfer from hollow sphere j to hollow sphere k (k=j+1), from equation 21 (μg/h)
R_{nn-1} = rate of mass transfer from solid sphere n to adjacent hollow sphere n-1, from equation 21 (μg/h).

$$v = \frac{4\pi}{3} \left(r_0^3 - r_i^3 \right) \qquad (37)$$

where v = volume of the hollow sphere (m^3)
r_o = outside radius of the hollow sphere (m)
r_i = inside radius of the hollow sphere (m).

6.5.5 Impermeable particles

The differential equations for impermeable particles are the same as equations 32 and 33.

6.5.6 Particles with a liquid film over an impermeable solid core

The differential equation for the liquid layer of particles is given by equation 38:

$$v_L \frac{dC_L}{dt} = R_{aL} \qquad (38)$$

where v_L = volume of the liquid film, from equation 37 (m^3)

61

C_L = SVOC concentration in the liquid film (μg/m^3)

t = time (h)

R_{aL} = rate of mass transfer from air to the liquid film, from equation 24 (μg/h).

$$v_L = \frac{4}{3} \pi \left(r_0^3 - r_1^3 \right) \tag{39}$$

where r_0 = radius of the particle (m)

r_1 = radius of the solid core (m).

6.5.7 Number concentration of airborne particles for a pulse release

The number concentration of airborne particles following a pulse release is given by equation 40 with initial conditions of $t = t_p$ and $N_p = N_{p0}$.

$$\frac{dN_p}{dt} = -(a + k_1) N_p \tag{40}$$

where t = elapsed time (h)

t_p = time for the pulse release (h)

N_p = number concentration at time t (counts/m^3)

a = ventilation rate (h^{-1})

k_1 = first-order deposition rate constant (h^{-1}).

6.6   Initial Conditions

The program sets the initial time at t = 0 hour. All other initial conditions are defined by the user. For example, the initial concentration in indoor air is defined in the <Conditions> page; the initial concentrations in different slices of the source are defined in form <Diffusional sources> in page <Sources>. After the model is compiled successfully, the user can view the initial conditions by clicking the <Inspect> speed button or select <Model>, <Inspect> from the main menu.

6.7   Unit Conversion

This program uses (μg/m^3) as the base unit for SVOC concentrations in all indoor media. The concentration units for the output data vary. This section presents the equations for unit conversion.

6.7.1 SVOC Concentration in sources and sinks

The program gives the SVOC concentration in sources and sinks in (μg/m^3), which can be converted to (μg/g) by equation 41:

$$C_{mw} = 10^{-6} \frac{C_{mv}}{\rho} \qquad (41)$$

where C_{mw} = SVOC concentration in source or sink material in (μg/g)
  C_{mv} = SVOC concentration in source or sink material in (μg/m^3)
  ρ = material density (g/cm^3).

Note that, if the output is the concentrations in different slices (see Section 6.3), the unit conversion must be done for individual slices because the slices have different thicknesses.

6.7.2 Particle-air partition coefficient

The particle-air partition coefficient is sometimes given in (m^3/μg). It can be converted to the dimensionless value by using equation 42:

$$K_p = 10^{12} K_p^{'} / \rho \qquad (42)$$

where K_p = particle-air partition coefficient (dimensionless)
  $K_p^{'}$ = particle-air partition coefficient (m^3/μg)
  ρ = particle density (g/cm^3).

6.7.3 Particle mass and number concentrations in air

For airborne particles in a given size bin or particle type, the conversion between the mass and number concentrations are done with equation 43:

$$N_p = 6 \times 10^6 \frac{M_p}{\pi \rho d^3} \qquad (43)$$

where N_p = particle number concentration (counts/m^3)

63

M_p = particle mass concentration (μg/m^3)
ρ = particle density (g/cm^3)
d = particle diameter (μm).

A unit converter is available for this purpose (<Tools> / <Unit conversion calculators> / <PM concentration in air>).

6.7.4 SVOC concentration for impermeable particles

The basic unit for SVOC sorption by impermeable particles is (μg/m^2). Equation 44 is used to convert the result to common unit (μg/g particles):

$$C_{pw} = \frac{6 C_{ps}}{d \rho}$$

(44)

where C_{pw} = SVOC concentration in impermeable particles (μg/g particle)
C_{ps} = SVOC concentration adsorbed by impermeable particles (μg/m^2 particle)
d = diameter of particles (μm)
ρ = density of particles (g/cm^3).

A unit converter is available for this purpose (<Tools> / <Unit conversion calculators> / <Impermeable PM>).

6.7.5 Dust loading versus number of dust

For a given size bin or particle type, equations 45 and 46 can be used to convert between dust loading and number of dust particles:

$$N_D = 6 \times 10^{12} \frac{L_D}{\pi \rho d^3}$$

(45)

$$L_D = \frac{N_D \pi \rho d^3}{6 \times 10^{12}}$$

(46)

where N_D = number of settled dust particles per unit area (counts/m^2)
L_D = dust loading (g/m^2)
ρ = density of dust particles (g/cm^3)
d = diameter of dust particles (μm).

To calculate the total number of dust particles in a given size bin or dust type, multiply N_D by the surface area. Note that both equations 45 and 46 are size-dependent. A unit converter is available for this purpose (<Tools> / <Unit conversion calculators> / <Dust loading vs number concentration>).

6.8   Miscellaneous calculations

6.8.1 Calculating the average particle-phase SVOC concentration in air in (µg/g particles)

The particle-phase SVOC concentration in air can be expressed in either (µg SVOC/g particles) or (µg SVOC/m^3 air). These concentration units require different calculation methods, as described below.

When expressed in (µg/g particles), the average particle-phase SVOC concentration is the *weighted mean* of all particles in room air, from equation 47:

$$\overline{C_{pw}} = \frac{\sum\limits_{i=1}^{n} N_{pi}\, w_i\, C_{wi}}{\sum\limits_{i=1}^{n} N_{pi}\, w_i} \tag{47}$$

$$w_i = \frac{10^{-12}}{6}\, \pi\, d_i^{\,3}\, \rho_i \tag{48}$$

where $\overline{C_{pw}}$ = average SVOC concentration in particle-phase in air (µg/g)

n = number of size bins/particle types
N_{pi} = number concentration for particles in size bin/particle type i (counts/m^3)
w_i = weight of one particle in size bin/particle type i, from equation 48 (g)
C_{wi} = SVOC concentration in particles in size bin/particle type i (µg/g)
d_i = diameter of particles in size bin/particle type i (µm)
ρ_i = density of particles in size bin/particle type i (g/cm^3).

6.8.2 Calculating the total particle-phase SVOC concentration in air in (µg/m^3 air)

When expressed in (µg/m^3 air), the total particle-phase SVOC concentration is the *weighted sum* of all particles in room air. Equation 49 takes into consideration both the number concentrations and particle-phase SVOC concentrations:

$$\overline{C_{pv}} = \sum_{i=1}^{n} N_{pi} C_{wi} w_i \qquad (49)$$

where $\overline{C_{pv}}$ = total particle-phase SVOC concentration in air ($\mu g/m^3$)

 n = number of size bins/particle types
 N_{pi} = number concentration for particles in size bin/particle type i (counts/m^3)
 w_i = weight of one particle in size bin/particle type i, from equation 48 (g)
 C_{wi} = SVOC concentration in particles in size bin/particle type i ($\mu g/g$).

6.8.3 Simulating episodic emission sources for airborne particles

Many indoor particle sources, such as vacuuming, cooking, smoking, and incense or candle burning, feature episodic particle generation. These sources are neither constant nor pulse injection. As a practical tactic, an episodic source of particles can be represented by a series of pulse injections (Guo, 2014). To do so, the number of pulse releases and the initial number concentration for each release must be determined. If the target number concentration during the episode is given, the amount of particles to be released in each pulse release can be estimated from equation 50 by using a root-trapping algorithm. It is assumed that the release times are equally spaced.

$$N_E = \frac{1}{t_1 - t_0} \sum_{i=1}^{n} \int_{t_{pi}}^{t_1} N_{pi}\, e^{-(a+k_1)(t-t_{pi})}\, dt \qquad (50)$$

where N_E = average number concentration during the episode (counts/m^3)

 t_1 = episode end time (h)
 t_0 = episode start time (h)
 n = number of pulse releases
 N_{pi} = initial number concentration for the ith pulse release (counts/m^3)
 a = ventilation rate (h^{-1})
 k_1 = first-order deposition rate constant (h^{-1})
 t = elapsed time (h)
 t_{pi} = release time for the ith pulse release (h).

This program contains a calculator (<Tools>, <Simulating episodic PM sources>) for determining the initial number concentration for each pulse release. To use this equation, the ventilation rate during the episode must remain constant. Figure 36 shows the simulated particle concentrations for the following hypothetical case:

Particle release event occurs between 2 and 3 elapsed hours

Particle diameter = 1 (μm)

Particle density = 1 (g/cm^3)

Particle mass concentration during the episode = 200 (μg/m^3), which is equivalent to a number concentration of 3.82×10^8 (counts/m^3) for the particles with the diameter and density mentioned above.

The event was represented by four consecutive pulse releases and the initial particle concentrations (Table 15) were calculated with the tool <Simulating episodic PM sources> under the <Tools> menu. Figure 36 shows the simulated number concentrations with demonstration model "PM-episodic.svoc". In most cases, the difference between the target and simulated number concentrations during the episode is within 2%.

Table 15. Representing an episodic particle-release event by four consecutive pulse releases

Release No.	Release time (h)	Initial concentration (counts/m^3)
1	2.00	4.634×10^8
2	2.25	1.528×10^8
3	2.50	1.528×10^8
4	2.75	1.528×10^8

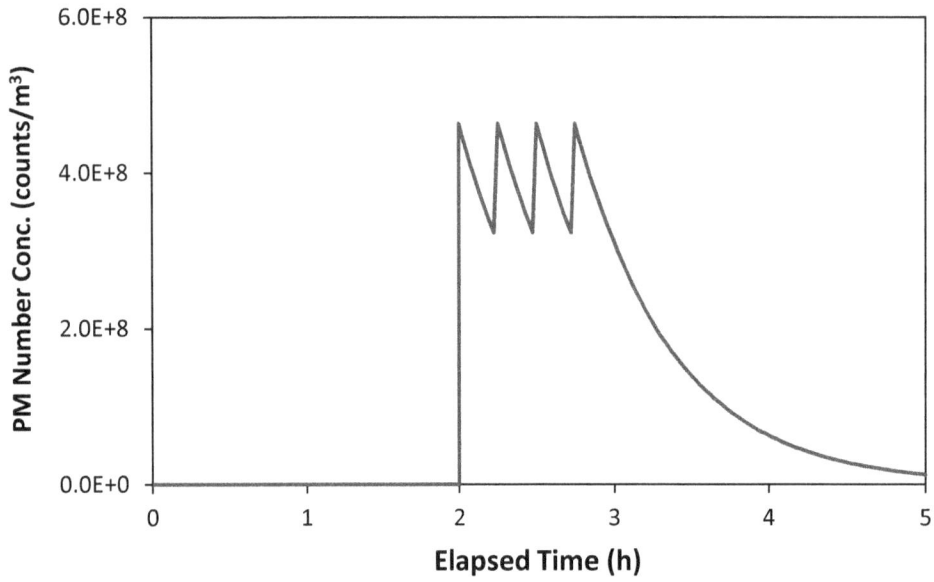

Figure 37. Simulated particle concentrations in room air due to an episodic particle-release event; created by demonstration model "PM-episodic.svoc".

6.8.4 Calculating the average SVOC concentration in settled dust

When there are multiple size bins/particle types for settled dust, the average SVOC concentration in the dust particles is the *weighted mean* of the concentrations of different size bins/particle types:

$$\overline{C_{dw}} = \frac{\sum_{i=1}^{n} n_{di}\, w_i\, C_{wi}}{\sum_{i=1}^{n} n_{di}\, w_i} \tag{51}$$

where $\overline{C_{dw}}$ = average SVOC concentration in settled dust (µg/g)

   n = number of size bins/particle types
   n_{di} = number of dust particles in size bin/particle type i (counts/m^3)
   w_i = weight of one dust particle in size bin/particle type i, from equation 48 (g)
   C_{wi} = SVOC concentration in dust particles of size bin/particle type i (µg/g).

68

6.9   SVOC migration from sources to settled dust due to direct contact

Program i-SVOC does not contain any models for SVOC transfer between sources and settled duct due to direct contact of the two phases. A simple method for estimating the *upper bound* of the SVOC concentration in the duct particles is given in Appendix A.

7. References

(Website accessibility was last verified on September 20, 2013)

Cheney, R. L. and Kincaid, D. (1980). Numerical Mathematics. Brooks/Cole Publishing Co., Pacific Grove, CA, pp 194-197.

Clausen, P. A., Hansen, V., Gunnarsen, L., Afshari, A., Wolkoff, P. (2004). Emission of di(2-ethylhexyl) phthalate from vinyl flooring into air and uptake in dust: emission and sorption experiments in FLEC and CLIMPAQ, *Environmental Science and Technology*, 38: 2531-2537.

Forthythe, G. E., Malcolm, M. A., and Moler, C. B. (1977). Computer Methods for Mathematical Computation. Prentice-Hall, Inc., Englewood Cliffs, NJ, pp 136-147.

Guo, Z. (2002). Review of indoor emission source models – Part 2. Parameter estimation, *Environmental Pollution*, 120: 551-564.

Guo, Z. (2005). *Program PARAMS User's Guide*, U.S. EPA, EPA/600/R-05/066, 32 pp.

Guo, Z., Liu, X., Krebs, K. A., Stinson, R. A., Nardin, J. A., Pope, R. H., Roache, N. F. (2011). *Laboratory study of polychlorinated biphenyl (PCB) contamination and mitigation in buildings — Part 1. Emissions from selected primary sources*, U.S. EPA, EPA/600/R-11/156, 107 pp.

Guo, Z., Liu, X., Krebs, K. A., Greenwell, D. J., Roache, N. F., Stinson, R. A., Nardin, J. A., and Pope, R. H. (2012). *Laboratory study of polychlorinated biphenyl (PCB) contamination and mitigation in buildings — Part 2. Transport from primary sources to building materials and settled dust*, U.S. EPA, EPA/600/R-11/156a, 166 pp.

Guo, Z. (2013). A framework for modeling non-steady state concentrations of semivolatile organiccompoundsindoors—I.Emissionsfromdiffusionalsourcesandsorptionbyinterior surfaces, *Indoor and Built Environment*, 22: 685–700.

Guo, Z. (2014). A framework for modeling non-steady state concentrations of semivolatile organic compounds indoors — II. Interactions with particulate matter, *Indoor and Built Environment*, 23:26-43.

Kumar, D. and Little, J. C. (2003a). Single-layer model to predict the source/sink behavior of diffusion controlled building materials. *Environmental Science & Technology*, 37:3821–3827.

Kumar, D. and Little, J. C. (2003b). Characterizing the source/sink behavior of double-layer building materials. *Atmospheric Environment*, 37: 5529–5537.

Li, W. G., and Davis, E. J. (1996). Aerosol evaporation in the transition regime, *Aerosol Science and Technology*, 25: 11-21.

Liu, Z., Ye, W., and Little, J. C. (2013). Predicting emissions of volatile and semivolatile organic compounds from building materials: a review, *Building and Environment*, 64: 7–25.

Nazaroff, W. W. (2004). Indoor particle dynamics, *Indoor Air*, 14 (Suppl. 7): 175–183.

Qian, J., Ferro, A. R., and Fowler, K.R. (2008), Estimating the resuspension rate and residence time of indoor particles, JAWMA, 58(4):502-516.

Schwope, A. D., Goydan, R., and Reid, R. C. (1990). Methods for assessing exposure to chemical substances. Volume 11: Methodology for estimating the migration of additives and impurities from polymeric materials. U.S. EPA, EPA 560/5-85/015, 148 pp.

Shi, S. and Zhao, B. (2012). Comparison of the predicted concentration of outdoor originated indoor polycyclic aromatic hydrocarbons between a kinetic partition model and a linear instantaneous model for gas-particle partition, *Atmospheric Environment*, 59:93-101.

Sparks, L. E. (2001). Indoor air quality modeling, in Spengler, J. D., Samet, J. M., and McCarthy, J. F. (eds.), *Indoor Air Quality Handbook*, McGraw-Hill, New York, NY, pp 58.1-58.23.

Takigami, H., Suzuki, G., Hirai, Y., and Sakai, S. (2008). Transfer of brominated flame retardants from components into dust inside television cabinets. Chemosphere, 73: 161–169.

Tichenor, B. A., Guo, Z., Dunn, J. E., Sparks, L. E., Mason, M. A. (1991). The interaction of vapor phase organic compounds with indoor sinks, *Indoor Air*, 1:23-35.

Van Loy, M. D., Lee, V. C., Gundel, L. A., Daisey, J. M., Sextro, R. G., Nazaroff, W. W. (1997). Dynamic behavior of semivolatile organic compounds in indoor air. 1. Nicotine in a stainless steel chamber, *Environmental Science and Technology*, 31:2554-2561.

Weschler, C. J., Salthammer, T., and Fromme, H. (2008). Partitioning of phthalates among the gas phase, airborne particles and settled dust in indoor environments, *Atmospheric Environment*, 42:1449–1460.

Webster, T. F., Harrad, S., Millette, J. R., Holbrook, R. D., Davis, J. M., Stapleton, H. M., Allen, J. G., McClean, M. D., Ibarra, C., Abdallah, M. A., and Covaci A. (2009). Identifying transfer mechanisms and sources of decabromodiphenyl ether (BDE 209) in indoor environments using environmental forensic microscopy, *Environmental Science & Technology*, 43(9):3067-3072.

Xu, Y. and Little, J. C. (2006). Predicting emissions of SVOCs from polymeric materials and their interaction with airborne particles, *Environmental Science and Technology*, 40: 456-461.

Appendix A

SVOC transfer to settled dust due to direct contact with a source

SVOC can migration from a source to settled dust by direct contact (Webster et al., 2009; Takigami et al., 2009). This mass transfer mechanism is especially important for chemicals with very low vapor pressures. The current version of i-SVOC does not include this mechanism because there are no published dynamic models that take into consideration both solid-solid and solid-air mass transfers. The two highly simplified models described below allow rough estimation of the SVOC concentration in settled dust that is in direct contact with a source. Both models ignore the SVOC transfer between the dust and air.

A.1 Equilibrium model

When equilibrium is reached between two solid materials that are in direct, firm contact, the distribution of the SVOC between the two phases is determined by equation A.1:

$$K_{dm} = \frac{C_d}{C_m} \tag{A.1}$$

where K_{dm} = solid-solid partition coefficient between the dust and source material (dimensionless)

C_d = equilibrium SVOC concentration in dust ($\mu g/m^3$)

C_m = equilibrium SVOC concentration in source material ($\mu g/m^3$).

Although few, if any, experimentally determined solid-solid partition coefficients are available for common building materials and dust, parameter K_{dm} can be estimated from the solid-air partition coefficients for the two phases (Kumar and Little, 2003b):

$$K_{dm} = \frac{K_{da}}{K_{ma}} \tag{A.2}$$

where K_{da} = solid-air partition coefficient for dust particles (dimensionless)

K_{ma} = solid-air partition coefficient for source material (dimensionless).

Thus, the equilibrium SVOC concentration in the dust can be estimated from equation A.3:

$$C_d = \frac{K_{da}}{K_{ma}} C_m \tag{A.3}$$

73

The values of C_d calculated from equation A.3 should be treated as the *upper bound* of the absorption because the calculation ignores the SVOC mass fluxes between the dust and room air and because it takes much longer time for large dust particles to reach equilibrium than for small particles.

The solid-solid partition coefficient, K_{dm}, can be either greater or smaller than 1, depending on the values of the two solid-air partition coefficients in Equation A.2. Takigami et al. (2009) observed that, in some cases, the concentrations of polybrominated diphenyl ethers (PBDEs) in settled dust collected from inside the television casing were higher than their concentrations in the sources (i.e., the casing and circuit boards).

A.2 Dynamic model

The rate of mass transfer between two solids that are in direct contact is given by Equations 7 through 11 in Section 6.4.1. For SVOC transfer from source to dust, Equation A4 applies:

$$R_{sd} = A H_d \left(K_{dm} C_m - C_d \right) \tag{A.4}$$

where R_{sd} = rate of mass transfer from source to dust particle (µg/h)
A = contact area (m^2)
H_d = overall mass transfer coefficient with respect to the dust phase (m/h)
K_{dm} = solid-solid partition coefficient, from equation A.2 (dimensionless)
C_m = concentration in the source (µg/m^3)
C_d = concentration in the dust (µg/m^3).

If the source is a large reservoir for the SVOC, which is the case for many SVOCs including phthalates and flame retardants, the sorption by the settled dust has little effect on the SVOC concentration in the source. Thus, we can assume that C_m is constant and, therefore, Equation A4 can be simplified because the overall mass transfer coefficient is no longer needed:

$$R_{sd} = A h_d \left(K_{dm} C_m - C_d \right) \tag{A.5}$$

where h_d = solid-phase mass transfer coefficients for dust particle (m/h).

74

Dust particles are in different shapes. To use Equation A.5, we further assume that the dust particle can be represented by a cube with a contact area of A and a height of ϑ. The solid-phase mass transfer coefficient, h_d, can then be estimated from Equation A.6:

$$h_d = \frac{2 D_d}{\theta} \tag{A.6}$$

where D_d = solid-phase diffusion coefficients for dust particle (m²/h)
ϑ = height of the dust particle "cube" (m).

Equation A.7 gives the mass balance for the dust cube:

$$v_D \frac{dC_d}{dt} = A h_d \left(K_{dm} C_m - C_d \right) \tag{A.7}$$

where v_d = volume of the dust cube (m³)
t = time (h).

Given the initial conditions of $t = 0$ and $C_d = 0$, Equation A.7 can be solved to give an exact solution for C_d (Equation A.8). Note that Equation A.8 contains neither v_D nor A.

$$C_d = K_{dm} C_m \left(1 - e^{-\frac{2 D_d}{\theta^2} t} \right) \tag{A.8}$$

where C_d = SVOC concentration in the dust (µg/m³)
K_{dm} = solid-solid partition coefficient, from Equation A.2 (dimensionless)
C_m = SVOC concentration in the source, treated as a constant (µg/m³)
D_d = solid-phase diffusion coefficient for dust particle (m²/h)
ϑ = height of the dust cube (m)
t = time (h).

To determine the value of ϑ in Equation A.8, the dust particle of whatever shape must be "converted" to an equivalent cube. This can be done in three steps: (1) calculate the volume of the dust particle (v_D); (2) define the contact area (A); and (3) calculate ϑ from $\vartheta = v_D/A$.

Figure A.1 was created with the following parameters:

$\vartheta = 1 \times 10^{-5}$, 2×10^{-5}, and 5×10^{-5} (m)
$K_{dm} = 0.5$
$D_d = 10^{-12}$ m²/h
$C_m = 10^{10}$ µg/m³ (equivalent to 10000 µg/g if density = 1 g/cm³)

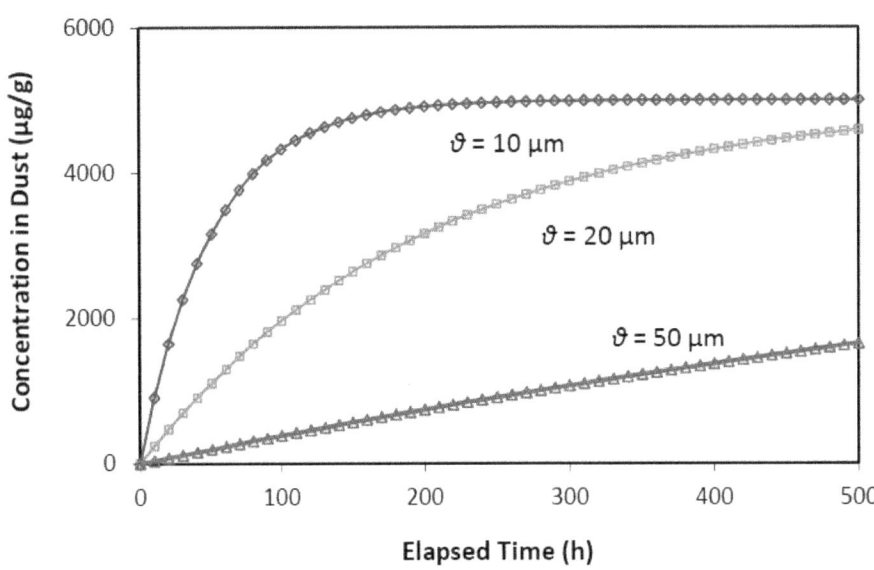

Figure A.1 SVOC migration from a constant source to settled dust due to direct contact of the two phases (assuming dust density = 1).

Appendix B
Parameter estimation for the dynamic Freundlich adsorption model

B.1. Purpose

This appendix describes a method for converting the equilibrium Freundlich adsorption model to a roughly equivalent, dynamic counterpart.

B.2 Equilibrium Freundlich model

The adsorption model proposed by Xu and Little (2006) assumes that a nonlinear, instantaneously reversible, Freundlich equilibrium relationship exists between the air and the surface, and that the SVOC concentration adsorbed on the surface at any time can be calculated from equation B.1:

$$C_s = K_f C_a^n \tag{B.1}$$

where C_s = concentration on sink surface ($\mu g/m^2$)
K_f = constant for a given chemical and sink material at a given temperature ($\mu g^{1-n} \, m^{3n-2}$)
C_a = concentration in air ($\mu g/m^3$)
n = constant for a given pair of chemical and sink material at a given temperature (dimensionless).

B.3 Dynamic Fruendlich model

The dynamic Freundlich adsorption model (Van Loyd et al., 1997) uses equations B.2 and B.3 to calculate the adsorption and desorption rates:

$$R_a = A f_a C_a^\alpha \tag{B.2}$$

$$R_d = A f_d C_s^\beta \tag{B.3}$$

where R_a = adsorption rate ($\mu g/h$)
R_d = desorption rate ($\mu g/h$)
A = area of sink surface (m^2)
f_a = nonlinear adsorption rate constant ($\mu g^{1-\alpha} \, m^{3\alpha-2} \, h^{-1}$)
f_d = nonlinear desorption rate constant ($\mu g^{1-\beta} \, m^{2\beta-2} \, h^{-1}$)
C_a = concentration in air ($\mu g/m^3$)

C_s = concentration on sink surface ($\mu g/m^2$)

α and β = dimensionless constants.

It is easy to prove that the equilibrium and dynamic Freundlich models are linked by equations B.4 and B.5:

$$K_f = \left(\frac{f_a}{f_d}\right)^{1/\beta} \tag{B.4}$$

$$n = \frac{\alpha}{\beta} \tag{B.5}$$

B.4 Estimating parameters α, β, f_a, and f_d from K_f and n

When the parameters for the equilibrium Freundlich model — K_f and n — are known, the parameters for the dynamic model — α, β, f_a, and f_d — can be estimated in three steps, as described below.

<u>Step 1</u>: Determine the values of α and β based on the value of n as follows:

If $n > 1$, set $\alpha = 1$, and $\beta = 1/n$

If $n \leq 1$, set $\alpha = n$, and $\beta = 1$

In either case, the ratio α/β is equal to n (see Equation B.5). It should be noted that the case with $n > 1$ is rare because, although equation B.1 is purely empirical, parameter n usually varies between 0 and 1.

<u>Step 2</u>: Assign a reasonably large value for f_a. As shown in equation B.4, it is the ratio f_a/f_d that really matters. The instantaneous equilibrium assumption implies that the adsorption rate is large. Thus, the value for f_a should be large. In the examples below, f_a is set to 10 ($\mu g^{1-\alpha}$ $m^{3\alpha-2}$ h^{-1}). An f_a value much greater than 10 is not recommended because it can cause numerical instability during the simulation.

<u>Step 3</u>: Calculate f_d from equation B.6, which is derived from equation B4:

$$f_d = \exp(\ln f_a - \beta \ln K_f) \tag{B.6}$$

B.5 Examples

Table B.1 summarizes the parameters for the examples described in Section 3.6. The simulation results are presented in Figures B.1 and B.2.

Table B.1. Sink parameters for DEHP emissions from vinyl flooring [a]

Model	Parameter Name	Parameter Values		Notes
		CLIMPAQ	FLEC	
Equilibrium	K_f	3800	6000	Xu & Little, 2006
	n	1.5	0.47	Xu & Little, 2006
Dynamic	α	1	0.47	From step 1
	β	0.67	1	From step 1
	f_a	10	10	From step 2
	f_d	0.040	0.0017	From step 3

[a] Other required parameters for creating the models are given in Table 6 in Section 3.6.

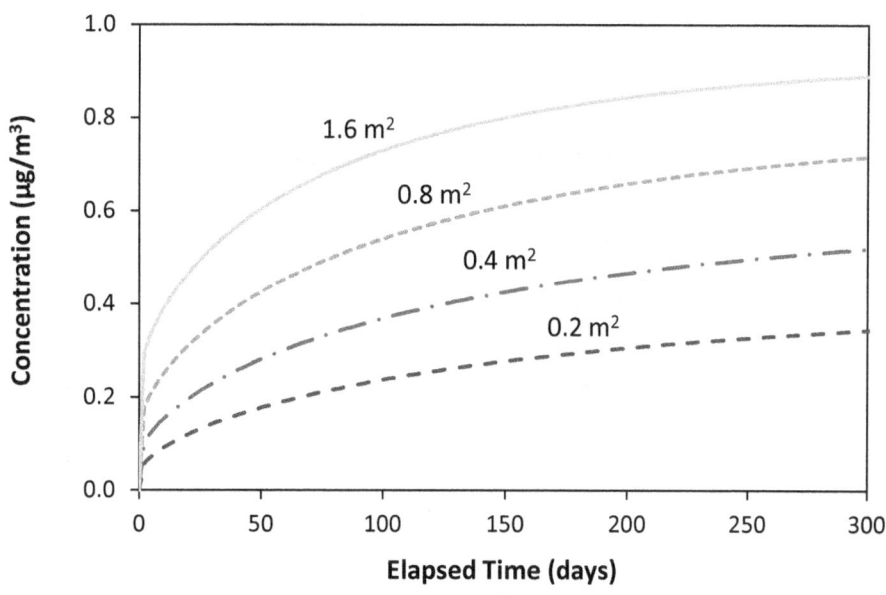

Figure B.1. Simulated DEHP concentrations in the CLIMPAQ chamber with the sink parameters given in Table B.1. Results are from four simulations.

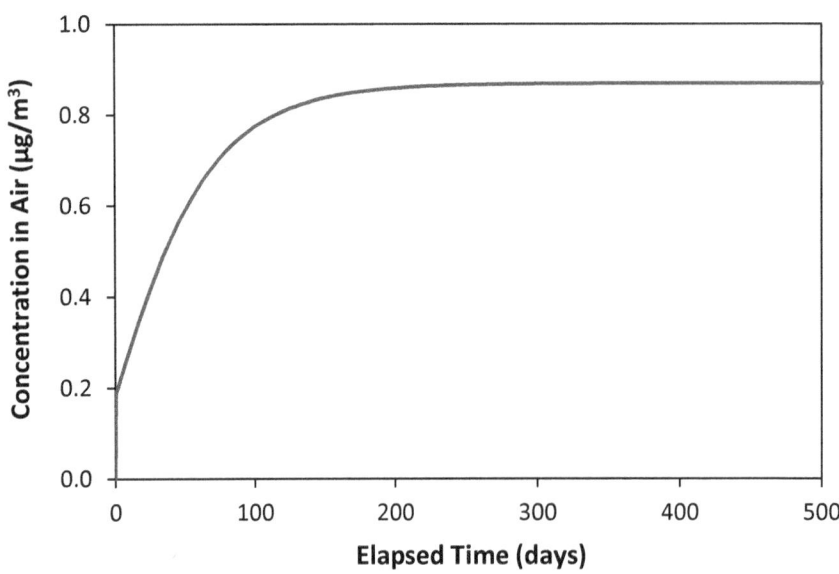

Figure B.2. Simulated DEHP concentrations in the FLEC chamber with the sink parameters given in Table B.1.

B.6 Caution

This method was tested with very limited data and, thus, should be used with caution.

www.ingramcontent.com/pod-product-compliance
Lightning Source LLC
Chambersburg PA
CBHW081552170526
45166CB00009B/2667